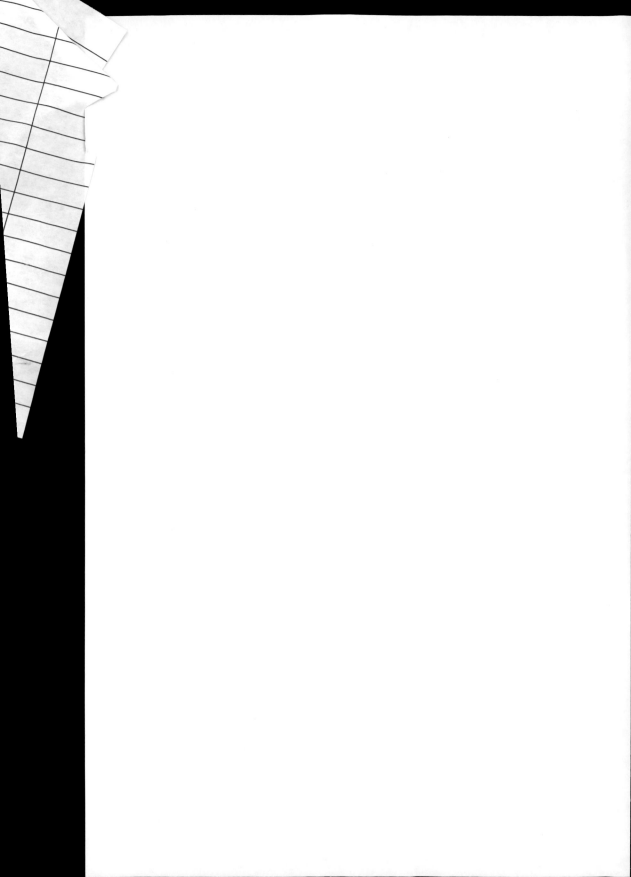

During the fifty-year period from 1936 to 1986 the modern agricultural revolution occurred, in which, for the first time, science was properly harnessed to the improvement in agricultural productivity. The authors quantify this improvement and identify the work of scientists which was seminal to the scientific and technological advances on which the revolution was founded. The topics covered include the advances in animal nutrition (in which the late Kenneth Blaxter was an acknowledged pioneer), animal and plant breeding, soil fertility, weed, pest and disease control, veterinary medicine, engineering (including innovations in tractor design by Harry Ferguson) and statistical measurement. In addition, this book describes how these innovations were integrated into the practical business of food production and discusses the importance of the Government in setting the scene for scientific advance.

From Dearth to Plenty
the modern revolution in food production

From Dearth to Plenty

the modern revolution in
food production

SIR KENNETH BLAXTER, FRS
NOEL ROBERTSON

CAMBRIDGE
UNIVERSITY PRESS

Published by the Press Syndicate of the University of Cambridge
The Pitt Building, Trumpington Street, Cambridge CB2 1RP
40 West 20th Street, New York, NY 10011-4211, USA
10 Stamford Road, Oakleigh, Melbourne 3166, Australia

© Cambridge University Press 1995

First published 1995

Printed in Great Britain at the University Press, Cambridge

A catalogue record for this book is available from the British Library

Library of Congress cataloguing in publication data

Blaxter, K.L. (Kenneth Lyon), Sir.
From dearth to plenty: the modern revolution in food production /
Sir Kenneth Blaxter and Noel Robertson.
    p.   cm.
Includes index.
ISBN 0 521 40322 7 (hc)
1. Agricultural innovations. 2. Agricultural innovations–Great Britain.
3. Agriculture. 4. Agriculture–Great Britain.
I. Robertson, Noel F. II. Title.
S494.5.I5B55 1995
338.1'6'09410904–dc20   94-25259CIP

ISBN 9 521 40322 7 hardback

VN

Contents

Preface

This book, which is concerned with the contributions of science to the progress of farming, particularly in the United Kingdom, was conceived in both pride and anger. Pride in the accomplishment of the many men and women, whether farmers or scientists, who transformed agriculture over the past half century, so that food could be provided from our resources of land and climate to meet the nutritional needs of the nation. Pride too in the fact that I have been privileged to know many of them. And anger? Anger that these gifted, often self-effacing people have not received the acclaim that is rightly their due. Jonathan Swift in 1727 recorded in Gulliver's Travels that the King of Brobdingnag '... gave it for his opinion, that whoever could make two ears of corn or two blades of grass, to grow upon a spot of ground where only one grew before, would deserve better of mankind, and do more essential service to his country, than the whole race of politicians put together'. Over 250 years ago, Swift recognised the accomplishment of those concerned with the then technology of farming and the contribution they made to the common weal. This we may well have forgotten in our age of affluence when hunger is no longer commonplace. It is wrong to have done so.

Although Swift rather denigrated politicians, there is little doubt that, but for a realisation by the political parties of the United Kingdom, during the period from the 1930s to the 1980s, that price and cost structures had to be favourable if farmers were to take up new methods and increase the production of food, the new technologies would not have been applied. The economic support to the farming industry was forthcoming and so too was the support for research, development and the essential task of encouragement of farmers to adopt new ways. For this reason the first two chapters of this book deal with the background to farming change in Britain over three centuries to provide the context in which innovation took place. The main part, however, deals with the discoveries that were made and the people who made them. The discoveries were many, varied, some of major, some of minor impact. It is completely impossible to detail all of them and so a selection has had to be made. The selection is a personal one and some readers may question the reasons for both inclusions and omissions. I apologise for the latter.

Obviously, I have had much help from those who were close to the contemporary scene of discovery and I indeed consulted them. I list them below and am extremely grateful to them for the unstinting help, through recollection, advice and by reference to the scientific literature, which they have given to me.

Kenneth Blaxter

Addendum.   When Kenneth Blaxter had completed rather more than half the book, he wrote to a friend that he must hurry to finish it. Sadly, he died shortly after. When Lady Blaxter asked me to complete the book I was glad to undertake the task, partly with the desire to serve a friend, partly because of a shared interest in the subject. We had often discussed how science should be carried into practice, and argued about it – sometimes heatedly! – in our days together on the Scottish Agricultural Development Council. Indeed, we had co-operated by encouraging staff from our respective institutions to work together. Before the writing of this book, however, it is doubtful if either of us realised just how complex the process of making science work for industry was. Here we have tried to give credit to the many people who had the seminal ideas which eventually changed the nature of the industry; at the same time we have noted the contributions of the people who saw the need for change and others who saw the potential of particular scientific ideas for eventual exploitation and development. All this we have tried to place in the context of the work of Government Research Institutes, University Departments and Colleges of Agriculture and of Veterinary Medicine, the Advisory and Development Services and the work of the agricultural chemical manufacturers and their research and development teams. At the same time we pay tribute to the civil servants in the Ministries and Departments who devised the fiscal policy against which change took place. We have not attempted particularly to demonstrate the success of the modern agricultural revolution in terms of production or productivity (efficiency), for its achievements unfold spontaneously as we proceed through the chapters.

The text focuses largely on British agriculture although sometimes work carried out elsewhere in the world is described in some detail because of its importance to agriculture in Britain. This is particularly true of the chapter on mineral deficiencies of animals (Chapter 9) where the solution of catastrophic problems in South Africa and Australia pointed the way for the solution of less dramatic but equally intransigent problems in Britain. Inevitably, too, although between us we have known and worked with a great many of the scientific protagonists, we have been able to demonstrate only a limited number of the undertakings and to mention only a few of the names involved. We apologise to those whose work has apparently gone unnoticed. Moreover, we have almost completely neglected the basic discoveries in the 'old' sciences of botany, biochemistry, chemistry, physics and zoology on which the recent work rests, but to incorporate these would have been a mammoth task. Nor have we attempted rigorously to establish the interactions between the agricultural advances and their effects on the environment, except where there is some immediate connection. Still less have we attempted to develop the very close relationship between many of the agricultural advances (in nutrition and reproductive physiology, for example) and the cognate advances in human medicine, although it is clear that they have been interdependent and that the agricultural advances have sometimes pre-dated those in medicine.

We have defined an agricultural revolution over the fifty-year period from 1936–1986. That revolution continues. The period after 1986 is distinguished particularly by the development of molecular biology, genetic engineering and the harnessing of computer technology to the instrumentation of biological research. It has brought with it a steep rise in costs, at a time when public funding is under pressure, and a bureaucratic and disciplined approach to the management of science. Research directors and university research scientists are forced, in their search for funding, to conform to a strategy imposed from without. In our discussion of the 1936–86 revolution, we are able to show the importance of strategic planning for the solution of industry's problems (the identification of needs). But we also show the importance of the freedom of directors and of applied workers to undertake speculative developments of new scientific ideas. This freedom to think the unthinkable and to develop the impractical was an important element in the success of the revolution. It sometimes seems as if the directors and research leaders of today, with every turn of the funding screw, are gradually having the last remnants of their freedom to innovate removed.

Information technology, molecular biology and genetic engineering are the sciences ripe for exploitation in the immediate future. But what lies beyond? Our success in the future must depend upon our capacity to identify and nurture new branches of science with relevance for agriculture and industry. To find these we need free spirits in the new areas of fundamental discovery and in the rapid development of these ideas as they come to hand.

Enough is said to demonstrate the complex interactions within the whole science activity in our revolutionary period; and to raise interesting questions about the process of discovery and development in a science-based industry. The analysis prompts comparison with the manufacturing industries, which are at present seeking to harness science to their purposes, indeed to suggest that they might look at agricultural development as a possible model. The important role of the state is setting fiscal parameters for the industry, the role of the generous science policy administered through the Research Councils and departments of state such as the Scottish Office and the Colonial Office, the important role of the Advisory and Development Services (sometimes effective, sometimes not) and the wonderful camaraderie and shared interest between scientists, advisors, developers, industrial scientists and farmers, which made the whole process go forward, are illustrated in these pages. It will be for others to draw on this material to establish where the process can be improved and to make the political judgements on the value for money of the investment in people and materials. No attempt has been made here to make judgements of the rightness of the actions described, which were driven by a need to improve and secure our food supplies. Certainly we both feel privileged to have taken part in this agricultural revolution and are more than content with the contributions of our colleagues.

Noel Robertson

Acknowledgements

In writing this book Sir Kenneth Blaxter consulted very widely among friends and colleagues to try to establish the important discoveries which had driven the modern agricultural revolution. The information received could have provided the basis of a more encyclopaedic work but it was Sir Kenneth's intention to make the material widely accessible and I have tried to follow him in this. Some of our friends have also read and corrected selected chapters. In the list below of people to whom we are indebted I have not separated those who provided information from those who read chapters, because in Sir Kenneth's case some of the information about who did what is not certain. Although the list is comprehensive there is the possibility that some of the people consulted have been overlooked by me in reading the files. For this I apologise. We have been greatly helped by those who read and corrected chapters and I have taken all their comments on board but I remain responsible for any errors or misinterpretations, both in the chapters written by Sir Kenneth and those that I wrote myself, and ask for their understanding in advance.

N. F. Robertson

Those who have helped us include the following:

Professor I.D. Aitken, Animal Diseases Research Association, Moredun, Edinburgh.
Mrs S.E. Allsopp, Rothamsted Experiment Station.
Professor D.G. Armstrong, Newcastle University.
Dr W.J.M. Black, Edinburgh School of Agriculture.
Professor I. Beneke, Iowa.
Dr D. Beever, Institute of Grassland and Animal Production, Hurley.
Professor P.M. Biggs, Royal Veterinary College.
Professor F.J. Bourne, Institute of Animal Health, Compton and Bristol University.
Professor P. Braude, National Institute of Research in Dairying, Reading.
Dr I. Bremner, Rowett Research Institute, Aberdeen.
Professor D.K. Britton, Wye College.

Mr K.N. Burns, Agriculture and Food Research Council.

Professor J.M.M. Cunningham, Scottish College of Agriculture and University of Glasgow.

Dr J.P.F. D'Mello, Edinburgh School of Agriculture.

Dr A.I. Donaldson, Institute of Animal Health, Pirbright.

Dr J.R. Finney, I.C.I. Agrochemicals.

Ms Wendy Foster, John Innes Institute, Norwich.

Professor Sir Leslie Fowden, Rothamsted Experiment Station.

Dr J. France, Institute of Grassland and Animal Production, Hurley.

Dr Richard Garnett, Monsanto plc.

Mr R. McD. Graham, Massey Ferguson.

Professor R.B. Heap, Institute of Animal Physiology, Babraham.

Professor W.G. Hill, University of Edinburgh.

Dr C.E. Hinks, University of Edinburgh.

Dr Robin Hawkey, Harper Adams Agricultural College.

Professor J.C. Holmes, Edinburgh School of Agriculture.

Dr Y. van der Honing, Netherlands.

Professor R.H.F. Hunter, Edinburgh.

Mr G.H. Jackson, Royal Agricultural Society of England.

Dr B.P.S. Khambay, Rothamsted Experiment Station.

Dr A.J. Kempster, Meat and Livestock Commission.

Dr B.R. Kerry, Institute of Arable Crops Research, Rothamsted.

Professor J.W.B. King, Animal Breeding Research Organisation, Edinburgh.

Mr C.A.W. Leng, Peeblesshire.

Professor T. Lewis, Institute of Arable Crops Research, Rothamsted.

Dr E.I. McDougall, Rowett Research Institute, Aberdeen.

Professor R.C.F. Macer, Consultant, Ely.

Miss Helen Martin, Rowett Research Institute, Aberdeen.

Mr A.O. Mathieson, Scottish College of Agriculture Veterinary Service, Edinburgh.

Professor J. Matthews, Institute of Engineering Research, Silsoe.

Professor J.T. Maxwell, Macaulay Institute of Land Use Research, Aberdeen.

Dr W. Meitz, Beltsville, Maryland, USA.

Professor K. Mellanby, Cambridge.

Dr C.F. Mills, Rowett Research Institute, Aberdeen.

Mr F.D.P. Moore, Howard Rotavators.

Dr J.H. Morgan, Institute of Animal Health, Compton.

Mr Phillip Needham, Agricultural Advisory and Development Service.

Mr Cedric Nielsen, Agricultural Advisory and Development Service.

Professor John Nix, Wye College.

Dr L.K. O'Connor, Milk Marketing Board.

Professor J.B. Owen, University of Wales, Bangor.

Mr T. Pennycott, Scottish College of Agriculture Veterinary Service, Auchencruive, Ayr.

Professor Sir Ralph Riley, Agriculture and Food Research Council.

Dr A.G. Roberts, I.C.I. Fertilisers.

Dr J.J. Robinson, Rowett Research Institute, Aberdeen.

Mr K.V. Runcie, Edinburgh School of Agriculture.

Dr R. Self, Rowett Research Institute, Aberdeen.

Professor N.W. Simmonds, University of Edinburgh.

Mr H.F. Simper, Farmer and Journalist.

Mr K. Simpson, Edinburgh School of Agriculture.

Mr W. Smith, Scottish College of Agriculture Veterinary Service, Craibstone, Aberdeen.

Professor Sir Colin Spedding, University of Reading.

Mr R.B. Speirs, Edinburgh School of Agriculture.

Dr J. Sutton, Institute of Grassland and Environmental Research, Hurley.

Professor P.N. Wilson, Edinburgh University.

Mr K.W. Winspear, Consulting Engineer.

PART ONE
The social, economic and political context of agricultural change

Throughout history and prehistory, the human race has had a major struggle to procure a safe and nutritionally adequate supply of food which is stable over time. There have been successes and failures in this struggle, and it continues, for the number of people in the world goes on growing while the land resource is fixed in amount. In this part, how change in agricultural practices came about in the past is described in the rather parochial context of United Kingdom farming. In addition it deals with the latest, and certainly the greatest, period of rapid change in agricultural production that has ever occurred in these islands.

1 Revolutions of the past

Revolutions in agriculture or in industry or in social custom differ from political revolutions, if only because they are not cataclysmic. They are not, as are many political revolutions, characterised by a growth of discontent which, suddenly focused, erupts swiftly to overthrow what has gone before. Rather, they are processes of slow change taking place over years and detected after such a lapse of time because they have inexorably altered what were accepted and traditional ways. Their origins are complex; their progress involves a variety of activities which interact and which vary in pace. Admittedly, and in common with political revolutions, charismatic individuals sometimes become identified with particular facets of the overall change, but seldom can these slow progressions be attributed solely to the zeal and endeavour of single protagonists or pioneers. The course of agricultural change in Britain over past centuries illustrates these complexities and helps to place the considerable changes in farming and the provision of food which have taken place in the past 60 years in a realistic perspective.

Agricultural revolutions in Britain

It is usual to call the considerable agricultural and agrarian changes that took place 200 years ago 'The agricultural revolution', the use of the word 'The' emphasising the importance that is usually attached to it and perhaps implying that it was unique. The revolution of the eighteenth century, although undoubtedly a major one, was not, however, unique; many changes of considerable magnitude had taken place in the centuries and millennia that preceded it. Hunting and gathering had given way to sedentary farming; animal power had displaced the power of human muscle; simple types of rotation of land between fallow and crop had been devised; and new concepts of land ownership had evolved.

Whether these considerable changes were a continuous progression or involved periods of rapid innovation interspersed with others of relative stagnation is not certain. Considering only the past thousand years it is probable that progress was rapid in the eleventh and thirteenth centuries. This period was followed by one in which the rate of change had slowed and it is probable that there was a fall in farm output that was not reversed until the sixteenth and seventeenth centuries. The evidence for the increase in farm output that occurred 800 years ago is largely based on estimates of the population of England, on the prices pertaining for wheat (for which there is a long statistical series), on interpretation of political events of the times, and on the surviving records of old estates and manors. The *Domesday Book* provides an estimate of the population of England in 1086 and the returns for the Poll Tax of 1377 one for that year. These figures suggest that the human population trebled in the course of two centuries. Subsequently it declined, only to resume its increase a century later. Food importation was not on any major scale, so it can be concluded that agricultural production must have increased considerably in accord with these population changes. Prices of wheat give support to the conclusion: they rose until about 1300, fell during the following two centuries and rose again during the sixteenth century. Anecdotal evidence about the technological changes that took place during the period from the Norman conquest to the time of the Black Death provides further support for the contention that an agrarian revolution of some magnitude took place during those times. Thus heavy ploughs were introduced from Europe and the two-field system of every crop being followed by a fallow was replaced by the three-field system, in which two grain crops were grown before the land was fallowed. Estimates of the increase of farm output that occurred in this earlier revolution can be made from yields of grain expressed as multiples of the amount sown. These suggest that yield per hectare probably increased from about 400 kg per hectare to about 700 kg during the period from the conquest to the middle of the fourteenth century.

This sparse evidence suggests that the first two centuries of the present millennium were a period of relatively rapid technical and structural change in farming. There is even sparser evidence of a similar revolution in farming taking place in the sixth century and by inference there must have been many others in the progression from primitive gathering and hunting methods of food procurement to the methods employed at the present time, which can be attributed to what may be termed the post-war agricultural revolution.

There is no agreed chronology of these revolutions. Even 'The' agricultural revolution of the eighteenth century causes difficulty. Thomson (1968) distinguished three revolutions between the late seventeenth century and the 1920s. The first or traditional revolution he identified with the demise of the three-field system and the adoption of rotations of crops. He thought that this was completed by 1815, to be followed by a second revolution characterised by the adoption of new inputs into farming in the form of fertilisers and imported feeding stuffs together with capital investment in buildings. The third revolution he regarded as beginning in 1914 with the adoption of labour-saving machinery.

The measurement of change

Given that an agricultural revolution is a period, often ill-defined, in which the rate of rural change has increased exceptionally, how can it best be measured? What are the attributes of progress that have been speeded up? We can distinguish three groups of such attributes.

The first group relates to the whole structure of rural society, involving land ownership, the division of responsibility for cultivation of the land, the location of the rural population, its subdivisions, its housing and its organisation. The decline in manorial and ecclesiastical influences in mediaeval times are examples of such structural changes. So too are the emergence of a class of landless agricultural labourers and the landlord-and-tenant system of the eighteenth century. Rural depopulation, the reduction in the number of village craftsmen and the centralisation of the industries that supply farming are more recent examples.

The second group relates to the primary purpose of agriculture, namely the provision of commodities and services which the community as a whole requires. Agriculture can be regarded as responding to the demands that society makes of it. Increases in food production in response to population growth, and increases in the proportion of the total food requirement that is met from home production, are obvious examples. So too are changes in the balance of commodities produced. The increase in the production of wheat in the early eighteenth century was at the expense of the rye and bere crops and was a response to a demand for wheaten bread. Similarly, the increase in the production of meat and milk in the late nineteenth and early twentieth centuries was a response to changes in the urban demand for food. These attributes are easily measured; they are simply the total outputs of the commodities concerned. How best to aggregate them to arrive at a single measure of output is more difficult. The usual way is to express them in monetary terms using contemporary prices. A more sophisticated way in which to express output is to state it relative to human demand. This leads to the concept of self-sufficiency, that is the proportion of human need that is met from home production.

The third group of attributes of agricultural progress relate to farming's ability to use the resources at its disposal. In this group the measure of progress is not production but *productivity* or efficiency. The resources that agriculture employs can be regarded as the classic ones of land, labour and capital. Accordingly progress can be monitored by estimating output of individual commodities per hectare of land, per person involved or per unit of resources employed from outwith the farm. It is these criteria of efficiency that give some measure of the technical progress of farming. How they might be combined to provide some single measure of productivity or to disentangle the various factors that collectively contribute to the increase in the yield of a crop or the output from an animal enterprise are difficult matters, and these have not been solved by economists in any agreed way.

Changed rural structures, increased output of commodities and improvements in productivity are the measures that can be used to assess agrarian and agricultural progress. Revolutions in agriculture can be discerned when the rate of change in all or some of these three groups of attributes substantially exceeds the slow improvement that is evident over millennia. It is interesting to apply these criteria to the events of the past three centuries.

The revolution of the eighteenth and nineteenth centuries

Using these criteria there is no doubt that a revolution in farming occurred to transform the agriculture of the middle of the seventeenth century to that of the late nineteenth and early twentieth century. As with the dating of earlier bursts of agricultural activity, there is considerable uncertainty about when the revolution began. Most historians agree that its impetus declined after the 1860s, the decade that marked the end of the era of 'high farming' in Britain and the beginning of the agricultural depression. Historians working in the early part of this century, notably Lord Ernle, while recognising that some agricultural improvement had taken place before, thought that progress really began on the accession of George III in 1760 and the advent of land enclosure by Act of Parliament. More recent studies give considerably greater weight to changes occurring early in the eighteenth century and indeed to those taking place in the late seventeenth century, thus placing the beginning of the revolution much earlier. Certainly land enclosures by private agreement had taken place long before those brought about by parliamentary Act. The adoption and field cultivation of the new crops of turnips and clover (which had been introduced from the Low Countries) began in the sixteenth and seventeenth centuries, while alternate husbandry – the process in which land was put down to grass for a few years before being cultivated again – was practised under some variants of the three-field system of Tudor times. A further difficulty about dating the beginning of the revolution is that its pace varied in different parts of the country. Changes were initially slow in Scotland but then accelerated such that by the nineteenth century Scottish farming was judged to be in advance of English practice. In the light of recent historical research the most acceptable dating of the revolution is probably that it occurred between 1700 and 1860. The earlier phases were dominated by structural changes and the later by technical ones as the new devices and methods and the improved animals and crops spread throughout the land to be modified and improved by those who used them. For convenience the revolution is called here the eighteenth-century revolution although it did continue well into the nineteenth century.

Dating of 'the' agricultural revolution in this way, although it can be criticised, has the merit of convenience because statistical information about farming is available for the last years of the seventeenth century and for the 1860s. Gregory King and Charles Davenant in 1698 and John Houghton in his *Collections* or broadsheets of the 1690s provide some estimates related to land, crops, people and livestock in England and Wales at the end of the seventeenth century. These can be supplemented (or at least roughly checked) by appeal to Sir William Petty's accounts of crop yields in Ireland in 1691. Statistical information for the end of the period is less difficult to find. In 1866, following a successful motion in the House of Commons, the Board of Trade began to collect estimates of the acreages of crops and numbers of livestock. Estimates of yields per unit area or of total output were not made officially until 1885 but, happily, Lawes and Gilbert at the Rothamsted Experiment Station began collecting and publishing information along these lines in 1852. The official statistics were not complete; initially some farmers refused to give information and informed guesses had to be made of the missing data. There is no doubt a lack of precision in the estimates that can be made of agricultural production and productivity both at the beginning and at the end of the revolution. Imprecision must thus apply to the differences between them which measure the crude overall rate of change, but this is hardly sufficient to vitiate them altogether.

Table 1 summarises these numerical estimates to show the changes in agriculture that occurred in England and Wales during the period from about 1700 to the 1860s. Arable land increased in area by about 60% and yields of wheat probably doubled. These estimates suggest that total output rose about threefold. The price of grain was much the same at the beginning and at the end of the period although it had risen massively during the Napoleonic wars. Human population increased by a factor of 3.7 during the period, which suggests that farm output did not quite keep pace with population growth. Such a conclusion is supported by the fact that Britain ceased to be an exporter of grain by the 1790s and subsequently became a net importer. It had exported grain during the Roman occupation, continued to do so during mediaeval times – except when disastrous harvests occurred – and was regarded by the French of the early eighteenth century as the granary of Europe. Britain was not to become a grain exporting country again until two centuries had elapsed, when, in the 1980s, it resumed an appreciable net export trade in grain.

The numbers of sheep and pigs increased during this revolution but the number of cattle fell. Part of this fall may well represent the continuing reduction in the number of draught oxen, which were kept until they were six or more years of age before they were killed. Contemporary descriptions of the

Table 1. *The changes that occurred in England and Wales during the agricultural revolution of the eighteenth and early nineteenth centuries*

The data for 1695–1700 derive from the estimates made by contemporary authors. The figures for family labour and for workers at that time are not very accurate since they are based on family size and ratios. The values for the agricultural population are based on the 1861 census of the population.

It will be noted that the product of the increase in the area of cultivated land and of estimated wheat yield per hectare is slightly in excess of the increase in the total population.

	in 1695–1700	in 1866	Ratio 1866/1700
Human population (millions)	5.8	21.4	3.7
Agricultural land area (million hectares)			
Cultivated area	8.5	10.0	1.2
Arable land	3.6	5.8	1.6
(Probable) fallow	(1.1)	0.3	0.3
(Probable) cropped area	(2.5)	5.5	2.2
Pasture and meadow	4.9	3.2	0.65
Agricultural population (thousands)			
Farmers and graziers	330	249	0.8
Family labour	500	112	0.2
Bailiffs and foremen	?	11	—
Agricultural workers	587	965	1.6
Livestock numbers (millions)			
Cattle	4.5	3.8	0.8
Sheep	12.0	16.8	1.4
Pigs	2.0	2.3	1.2
Wheat yield (kg h^{-1})	900	2070	2.3

cattle killed for meat suggests that they had changed remarkably from the scrawny, undernourished animals characteristic of the beginning of the period to beasts with considerable body fat and higher carcase weights at the end. Estimates made by Trow-Smith suggest that milk yields of cows in the late seventeenth century were about 200 gallons per year (about 900 litres per year). From sparse figures given by Youatt in the mid-nineteenth century, milk yields had probably doubled by this time.

Changes in the numbers of people who were directly engaged in farming are more difficult to ascertain. The data in Table 1 are somewhat suspect because the number of farm workers is based on the assumption that the ratio of landless workers to farmers was 1.78 : 1. Even so the figures suggest that the number of farmers fell but the number of men and women they employed increased. The total number of people engaged in agriculture did not increase greatly, if at all, and certainly did not change at the rate at which the total population of the country increased. What information is available does not support the idea that this revolution in farming resulted in a massive release of labour from the land to fuel the industrial revolution, except in the sense that farming's labour force did not greatly increase although that of the towns did so.

The new technologies of the eighteenth-century revolution

What may be termed the heroic view of the eighteenth-century revolution singles out a few of the innovations in farming and identifies them with individuals. The Norfolk four-course rotation, with the cropping sequence, wheat–turnips–barley–clover, is thus identified with Viscount Townshend; the light horse-drawn plough to replace the cumbersome implement of the sixteenth century is regarded as being due to the genius of James Small. Jethro Tull is always singled out as the man whose invention of the seed drill to replace the broadcasting of crops and so permit inter-row cultivation did most to further technical advance. With livestock, Robert Bakewell's selective breeding of sheep to produce the new Leicester breed and the similar work of the Colling brothers in breeding improved shorthorn cattle were similarly regarded as lone, pioneering and heroic ventures which sparked the technical aspects of the revolution.

More recent scholarship, although it does not diminish the importance of these technical advances in contributing to the increase in production and productivity of farming, shows that they can hardly be attributed to these individuals alone. The four-course rotation was in operation long before 'Turnip' Townshend popularised it. Small's light plough originated in Flanders and owed much to the experiences of Arbuthnot. Tull's seed drill had been anticipated in the seventeenth century and his turnip husbandry was not new in 1731 when he published his *New Horse-Hoing Husbandry*. Even the writers of the time, such as William Marshall, thought that Bakewell was only one of many contemporaries who were attempting to improve livestock by selection of superior animals and breeding from them. There were, of course, exceptions; the erosion of the reputation of the agricultural heroes has not been absolute. There is no doubt, for example, that the invention of the cast iron, self-sharpening ploughshare was solely due to Robert Ransome.

Two conclusions can be drawn about the relation between technical advance and the course of the eighteenth-century revolution. The first is that technical innovation was a continuous process of invention and modification, and that many landowners, farmers and craftsmen took part, some much more effectively than others. Additionally, through the efforts of agricultural writers and popular lecturers, notably Arthur Young, the accomplishment of these leaders reached the attention of a wider audience. The interest evoked by these lectures and writings, and the growth of local agricultural societies, where opinions about best courses of farming action were debated, suggest there was a real enthusiasm for what was called 'agricultural improvement' during that time.

The second conclusion that can be drawn follows upon this role of the agricultural writers and popularisers. There was a considerable lag between the emergence of ideas about tools and methods and their widespread adoption. Seed drills were not common in East Anglia until the nineteenth century although they had been invented in the seventeenth century and refined by Tull in the early eighteenth. It may be that some who date the beginning of the revolution to the middle of the seventeenth century place too much emphasis on when the technical means of effecting increases in production, productivity and agrarian structures were first devised rather than when they were employed on a meaningful scale.

Practice with science

What is remarkable about the technical advances in farming that were made during the eighteenth-century revolution is that they were the achievements of practical men working on, or close to, the land. They were improvements in husbandry, based on simple observation followed by practical test, rather than developments arising from analysis of the principles involved or from knowledge of how crops and animals grow. This is understandable. The science of the time was woefully inadequate and quite insufficient to provide any positive lead. Early eighteenth-century chemistry was dominated by the phlogiston theory, and the elements making up matter were still thought to be the classic ones: earth, air, fire and water. Mechanics was more advanced but biology was still very much a descriptive study without any cohesive structure. Even so, many of the eighteenth-century writers on agricultural improvement averred that their findings were scientific ones. Science appears to have been a term employed to confer an air of respectability on what these practical men had found to work.

The high esteem in which science was held persisted into the nineteenth century, when practical inventiveness was at its height. The Royal Agricultural Society of England, founded in 1838 and receiving its first Royal Charter in 1840, reflects this attitude. The Society, after much deliberation, took as its motto, 'Practice with Science'. Yet the records of the Society during its first 20 years of existence show that there was little of the union of farm practice and scientific investigation which that motto implies. Few of the developments in farming had derived from independent discoveries in chemistry, physics or biology. Rather, as in the previous century, the advances were justified through reference, often forced, to the new science which was then emerging. Indeed there is evidence of a disillusionment. Some questioned whether science had really resulted in any improvement in farming at all. Searching through the early minute books of the Royal Agricultural Society, Lynette Peel found that in 1842 its Council states that they were 'most anxious, at the present moment, to guard their members against the opposite evil of the undue and arbitrary application of mere unaided and theoretical science to the practice of agriculture'.

Despite these contemporary misgivings, science was at that time beginning to be employed in a positive way to improve farming practice. The outstanding example of this was the work of John Bennet Lawes and the founding of Rothamsted Experiment Station. Lawes, perhaps influenced by Justus von Liebig in Germany, commenced field trials and experiments in pots to elucidate the value of ground bones as a fertiliser. Liebig had suggested that dissolving bones in acid would increase their value as fertiliser. Lawes took this suggestion much further to show that bones and mineral phosphates could be treated with sulphuric acid to make an excellent fertiliser: 'superphosphate'. He patented his work in 1842, established a factory at Deptford Green in London, and made a considerable fortune from the sale of his 'patent manure'. He developed this investigational approach at Rothamsted: he appointed Joseph Gilbert to help him, and together these two men showed how science could be allied to the age-old art of farming.

Technical progress in the nineteenth century was immense. New methods of draining land using preformed clay pipes, new machines for sowing, cultivating and harvesting crops, and new buildings to accommodate stock and store produce were devised, modified and developed at an increasing pace during the latter part of the revolution. A new appreciation of the value and specific attributes of traditional manures such as soot, refuse from towns and slaughterhouse wastes arose; new sources of plant nutrients such as Peruvian guano, nitrate of soda and the by-products of coal gas manufacture were introduced. Steam power was applied to farming in the late 1850s both to run barn machinery and for ploughing. Constructive breeding of livestock accelerated, with the emergence of registers of animals in the form of herdbooks. Protein-rich feeding stuffs were derived from the oil seeds grown in the Colonial Dependencies (often crushed and extracted in Britain) and much was attempted to mitigate the effects of disease in both animals and crop plants. The later phases of the revolution in agriculture were years of high technical accomplishment and led to the emergence of the so-called 'era of high farming' of the 1840s to 1860s.

Looking at production, productivity, rural conditions and the technologies applied to farming at the end of the seventeenth century and in 1866, we see that a real revolution occurred in Britain as a whole. Simply looking at the beginning and end of the period conceals the fact that progress first accelerated, then stabilised and finally declined, and that it was not of equal pace in all parts of the country. The response of farmers and landowners varied from time to time and from place to place. The major determinants of the temporal change were the prices paid for farm produce and the farmers' expectation of reasonable returns. Sequences of years with unfavourable weather and the dislocations caused by wars resulted in considerable vagaries in prices and consequent variation in output. Unfavourable weather not only made field operations difficult or impossible but encouraged disease in both crops and stock. The frequent references in contemporary literature to outbreaks of 'the rot' (liver fluke, a parasitic disease) in sheep and 'murrain' (rinderpest) in cattle, to blight or smut in grain, and to the potato famine caused by potato blight, are indicative of the problems that wet seasons brought. As far as the spatial variation of progress is concerned, it has to be remembered that farming conditions vary considerably from one part of the country to another. Improvements that were successful in one part of the kingdom had to be modified before they were acceptable in others. Communications within Britain were poor, particularly before the advent of the railways and the improvement of roads. The exhortations of the popularisers to adopt new methods did not reach many and were seldom disseminated by those who received them.

The revolution was uneven, progressing and then regressing for a while; its principles were adopted with alacrity by some but viewed with conservative suspicion by others. The driving force was an economic one, with expectations of financial gain by the farmers and landowners the goals. To achieve these ends, considerable ingenuity – and indeed genius – was displayed in devising new ways of managing, reclaiming and cultivating the land, in organising its cropping, and in improving its livestock. The outcome was a vast change in agrarian structures, an increase in food production and the realisation that, to take advantage of economic opportunities, new technologies had to be adopted.

From depression to depression

In that we have defined an agricultural revolution as an acceleration of progress in food production above a background of slow change, it is informative to consider what happened after the end of the eighteenth-century revolution. The period from the 1860s to the 1930s is indeed the baseline from which one can judge the extent of the revolution in farming that has taken place between the 1930s and 1980s. What follows is an account of the fate of farming in Britain from the end of the era of high farming.

Farm production was at its peak in the 1850s and 1860s. More and more farmers had by then accepted the new technologies; their expectations were high and so too were the prices of the commodities that they had to sell. Lord Ernle, when writing in 1912 of this period, said 'Crops reached limits which production has never since exceeded, and probably, so far as anything certain can be predicted of the unknown, never will exceed'. Lord Ernle was wrong in his prediction. The limits were to be exceeded, but only after the lapse of a century.

The years of high farming were followed by the great agricultural depression which began in the mid-1870s and reached an appalling depth in the mid-1880s in terms of financial hardship for the farming community. The depression was followed by a slow recovery until the First World War. During that war, agricultural production was encouraged by guaranteed prices for farm produce under the Corn Production Act of 1917. Output rose in consequence, but the repeal of the Act in 1921 heralded a further decline in farm output exacerbated by the financial problems of the early 1930s.

The agricultural depression of the 1930s was just as severe as that of Victorian times. The land was neglected, hedges were untrimmed, ditches were not cleaned, cottages were allowed to decay and cropping was abandoned in many areas that were marginal for arable farming. Furthermore, there was an air of despondency about the farming community and an acceptance of impoverishment.

The decline in production from 1860 to 1938

We can compare the state of farming at the end of the period of high farming and that at the onset of the Second World War. The statistical series on which to base such estimates is not complete in all respects; happily, however, the information collected by Lawes and Gilbert is of considerable help in that it provides an overlap with the data subsequently collected by the government.

Lawes and Gilbert collected information on which to estimate the wheat output of the United Kingdom (which then included the whole of Ireland and not just the North). These show that in the eight years ending in 1860, home production of wheat accounted for 73% of the total supply and imports for 27%. By 1891 their figures show that the proportions had reversed; 73% was imported and only 23% was home-produced. The depression of the 1930s was as severe; home production of wheat again met only 23% of home demand. Unfortunately we do not have statistical information on the production of other crops or of stock in 1860 which could be used to assess the change that occurred in output of these commodities when high farming came to an end. However, we do have records of acreages of crops, numbers of stock and prices. Comparison of these for the years 1866 and 1938 give some indication of farming's plight as depression followed depression. These statistics are given in Table 2; they show that the area of land cropped with cereals fell by 40%, land

Table 2. *The area of crops, the numbers of farm animals and the prices of commodities in Great Britain in 1866 and in 1938*

	1866	1938	ratio, 1938/1866
Crop area (million hectares)			
Wheat	1.35	0.78	0.57
Barley	0.90	0.40	0.44
Oats	1.12	0.85	0.76
Rye	0.02	0.01	0.50
Total cereals (million hectares)	3.40	2.03	0.60
Peas and beans	0.26	0.07	0.27
Potatoes	0.20	0.25	1.25
Fodder root crops	0.98	0.38	0.39
Sugar beet	—	0.14	0.14
Temporary grassland	1.49	1.36	0.90
Permanent grassland	4.51	7.05	1.56
Total non-cereals	7.44	9.25	1.24
Total agricultural land, million hectares	10.84	11.28	1.04
Livestock numbers (millions)			
Cattle	4.79	8.03	1.68
Sheep	22.05	25.88	1.17
Pigs	2.48	3.82	1.54
Cereal prices (£/tonne)			
Wheat	11.7	6.8	0.58
Barley	10.5	10.2	0.97
Oats	8.8	6.9	0.78

cropped with turnips, swedes and mangolds by 62% and temporary grassland – the central feature of alternate husbandry – by 10%. Prices for cereals were less in 1938 than they had been in 1866. Livestock numbers increased and so did the area of land under permanent pasture. Much of this consisted of abandoned arable land, which was simply allowed to tumble down to grass.

The drastic reduction in grain production commenced in the 1870s when springs were cold and wet and harvests were poor. Shortages of grain led to higher prices for a while but then prices fell, discouraging corn production. The reason for this fall in price was the importation of cheap grain from the new lands of North America and Australia. There was no protection of the U.K. farming community from overseas competition: the Corn Laws had been repealed in 1846 and the country was wedded inexorably to free trade. The reaction of farmers was to turn to livestock enterprises. Quoting the old adage about prices: 'Down corn, up horn'. The demands of the growing number of townsfolk engaged in manufacturing industry for meat and dairy products held promise of new market opportunities, and grain for feeding to livestock was cheap. A reorientation of farming, with greater emphasis on livestock, worked for a while. This type of farming suffered a severe blow when in the 1880s refrigerated transport made it possible to import carcase meat to add to the increasing imports of cheese and butter. Considerable inroads were made on the home market by frozen and chilled meat from Argentina and, later, New Zealand. Imports of food paid for by the highly successful industrial manufacturing industries created economic circumstances that thrust the farming industry deeper into recession. The censuses of the United Kingdom show that the human population almost doubled from 1871 to 1931, when it reached 46 million. Its own farms did not provide the food for these people: the farms produced less at the end of the period than they did at the beginning.

The stagnation of productivity in the period 1860–1938

Most farmers during the period of the two depressions reduced the inputs into their enterprises. Some, however, took advantage of the new technologies that were emerging to devise new farming strategies to overcome the problems of low prices and high costs. Despite the overall fall in farm output, technical progress and inventiveness was considerable. The new steel industry, based on the Bessemer converter, produced a phosphate-rich waste product which, when finely ground to form basic slag, proved an excellent fertiliser for grassland. The reaper–binder, with its clever knotting device, provided sheaves at low labour cost. The milking machine was developed, and during the First World War the tractor was introduced to replace the horse. Originating in Germany, new and precise methods of designing the rations for all classes of farm livestock were introduced to effect an economy in feeding them. Sugar beet was introduced to replace the turnip and mangold crops which had been such an important feature of the Norfolk four-course rotation. This crop still enabled inter-row cultivation and weed control but the tops of the beet were a good feed for cattle and sheep while the roots – for sugar – provided additional income from prices buffered against market fluctuations.

Despite these and many other technical advances, and despite increasing scientific knowledge about the biology of plants and animals, the yields of crops per hectare of land – which are a good index of the technical efficiency of farming – showed negligible increases during the whole of the 70 year period. Again we are indebted to Lawes and Gilbert for information about yields in the early part of the period until, in 1885, official statistics were produced. Lawes

Table 3. *The mean yields of crops in 1886 and in 1936 showing the very small increase in yield that occurred during the depressions of British agriculture*

The values for 1886 are the mean of the three years 1885–7 and those for 1936 the mean of the three years 1935–7.

	Yield (t ha^{-1})		Ratio
	1886	1936	1936/1886
Wheat	2.08	2.16	1.03
Barley	1.97	2.00	1.02
Oats	1.62	2.02	1.25
Potatoes	15.1	16.5	1.09
Turnips and swedes	32.8	33.7	1.04
Hay	3.28	2.89	0.88

and Gilbert's figures show that the mean yield of wheat during the three-year periods centred on 1866 and 1885 were 2.07 and 2.01 t ha^{-1}, respectively (30.8 and 30.1 bushels per acre). Their figures thus show no increase in yield during this period. The official estimate of wheat yield in 1885 agrees well with that given by the Rothamsted scientists. Subsequently, from 1885 to the middle of the 1930s, yield of wheat per hectare showed little increase. The average yield during the three years centred on 1936 was 2.15 t ha^{-1} and thus the increase in wheat yield during 70 years was only 3%. Table 3 summarises changes in yields of crops in the half century from 1886 to 1936. They too show small or negligible increases. The contrast with the doubling in yield during the agricultural revolution of the eighteenth century is striking.

Another index of farming's efficiency relates to its use of manpower. One of the economies that farmers made during the decades of depression was to reduce the number of people they employed. The censuses of the population made at ten-yearly intervals are slightly complicated by changes in the ways in which people were classified in terms of their employment. They nevertheless show such a reduction. In 1861, during the times of farm prosperity, there were 312 000 farmers and 1.6 million farm workers. By 1931 the number of farmers had declined by 6% to 294 000. The number of farm workers, however, had halved to 800 000. Since 1921 the agricultural departments of the government have collected statistics on the number of farm workers every year. These show a relentless decline during the 1920s and 1930s and they are parallelled by a decline in the number of farm horses. When a man left the land his horse went with him. These statistics reflect the decline in arable cultivation as well as the replacement of the horse by the tractor. With a static or slightly declining production, the considerable decline in manpower must imply some increase in the efficiency with which farmers used their resource of labour.

The reaction of Government

The plight of the farming community during the depressions did not escape the notice of the government. This was understandable because in the 1880s more than half the members of the House of Commons were landowners. Much minor legislation was enacted and several major investigations were mounted to examine the state of the industry. The Duke of Richmond's Commission of 1879 revealed the widespread distress of the time and predicted that it was unlikely there would be any increase in prices to ameliorate the hardship. The Royal Commission of 1893–5 painted an even gloomier picture. It described the large areas of farm land which had passed out of production: at the same time, however, and in a more positive way, it drew attention to the few farmers who, appreciating the changed circumstances, had used new technologies to devise systems of farming that appeared viable. No solution to the widespread difficulties of the many was presented, other than to abjure landlords to reduce rents. The concept of free trade prevailed and was regarded as sacrosanct. With a brief departure from this concept, when unrestricted U-boat warfare during the First World War curtailed food supplies, this same attitude prevailed in the years that followed.

There were, however, some farsighted actions on the part of the government which were of major importance in providing a base for future expansion. The first of these was an Act of 1890, which gave County Councils the power to provide technical education in agriculture. The funds were known as 'the whisky money' since they were the residues of sums raised from the taxation of alcohol and which had hitherto been earmarked for the compensation of displaced or redundant publicans. As a result lecture courses and classes in agriculture were mounted, and colleges and university departments of agriculture were established, all with the purpose of improving the technical proficiency of the practitioners of agriculture. The same whisky money included a small modicum for the support of research. The poverty of that sum is evident from the fact that in 1909 only £425 was allocated for experiments and investigational work for the whole of Britain. It was at least a beginning.

The second advance was due to the efforts of David Lloyd George, the Chancellor of the Exchequer. He successfully introduced the Development and Road Improvement Funds Act of 1909, which was designed to develop the countryside and the coastal regions of the country. A very small part of the function of the Development Commission which was to administer these funds was to nurture appropriate research work. The funds set aside were originally £2 500 000 per annum. Later additional funds were voted; although, like all funds to promote research, these were never wholly sufficient, they provided a base on which the research structure in agriculture was later developed.

A third advance was certainly the establishment of the Agricultural Research Council in 1931. Its purpose was to advise the agricultural departments of Government about how they should support agricultural research. In some respects this was a tidying-up operation, designed to bring civil support of farming research into the same structure as that for medicine and industrial research. As the years passed, however, the Agricultural Research Council came to play a major part in the forging of the new technologies which were to transform farming. How this took place is described in more detail in Chapter 13.

The promotion of education by the 1890 Act, the identification of research as worthy of funding by that of 1909 and the establishment of the Agricultural Research Council as a central authority to coordinate research laid the foundation for the application of science to the age-old arts of husbandry. Men and women were trained to undertake investigations and the more progressive farmers were introduced to new knowledge about the soils, plants and animals on which their livelihood depended. As will be evident in later chapters, many of the advances made during the revolution of the 1930s to the 1980s had their origin in discoveries which were made much earlier as a consequence of the passing of these two Acts and the creation by Charter of the Agricultural Research Council. The discoveries themselves did not result in revolution any more than the invention of the seed drill or the four-course rotation, 200 years before, resulted in the revolution of the eighteenth century. Other factors relating to the farmer's appreciation of the promise of future prosperity were involved before there could be change and the emerging technologies could be adopted and modified to meet specific needs and to provide future economic security. The inventions that were characteristic of the recent revolution were not born of necessity; their adoption, however, certainly was.

The United Kingdom was not alone in this endeavour; throughout Europe and America similar initiatives were being taken and in Britain's colonial dependencies the structures developed in the home country were being adapted to meet different problems. It will be evident later that this international investment was highly relevant to the more parochial context of British farming.

Selected references and further reading
Statistical sources

Central Statistical Office (1939–1990). Annual Abstract of Statistics. Her Majesty's Stationery Office, London.

Houghton, J. (1691–1703). *A Collection of Letters for the Improvement of Husbandry and Trade.* Richard Taylor, London.

King, G. (1696). *Natural and Political Observations and Conclusions upon the State and Condition of England*. London.

Lawes, J.B. and Gilbert, J.H. (1893). Home produce, imports, consumption and price of wheat over 40 harvest years 1852–3 – 1891–92. *Journal of the Royal Agricultural Society of England*, Third Series **4**, 77–133.

Marks, H.F. (1989). *A Hundred Years of British Food and Farming: A Statistical Survey*, ed. D.K. Britton. Taylor and Francis, London.

Petty, W. (1691). *The Political Anatomy of Ireland*. D. Brown and W. Rogers, London.

Whitworth, C. (1761). *Political and Commercial Works of Charles Davenant*. London.

Historical sources

Command 2644 (1905). *Report of the Royal Commission on the Supply of Food in Time of War*. Her Majesty's Stationery Office, London.

Chambers, J.D. and Mingay, G.E. (1966). *The Agricultural Revolution 1750-1880*. B.T. Batsford, London.

Ernle, Lord (1936). *English Farming, Past and Present*. New 5th Edn by Sir Daniel Hall; New Imprint 1941. Longmans Green and Co., London.

Grigg, D. (1982). *The Dynamics of Agricultural Change*. Longmans, London.

Kerridge, E. (1967). *The Agricultural Revolution*. George Allen and Unwin, London.

Orwin, C.S. and Whetham, E.H. (1964). *History of British Agriculture*. Longmans Green and Co., London.

Peel, L.J. (1976). Practice with Science–the first twenty years. *Journal of the Royal Agricultural Society of England* **137**, 9–18.

Russell, E.J. (1966). *A History of Agricultural Science in Great Britain*. Routledge and Paul, London.

Thomson, F.M.L. (1968). The second agricultural revolution. *Economic History Review* **21**, 62–77.

Trow-Smith, R. (1959). *A History of British Livestock Husbandry, 1700–1900*. Routledge and Paul, London.

Tull, J. (1733). *The New Horse-Hoing Husbandry*. Published for the author. London.

Whetham, E.H. (1978). *The Agrarian History of England and Wales, vol. VIII*. Cambridge University Press.

2 *The modern revolution, its origins and accomplishments*

Dating the revolution

No doubt future historians will argue about the precise dating of the beginning and the end of the modern revolution in agriculture, in much the same way as they have argued about the dates of the agricultural revolutions of the eighteenth century and earlier. These future arguments will revolve around questions about what criteria to use to judge the extent of change. If government policies designed to support agriculture are considered, then it can be argued that the revolution began either in 1931 with the establishment of marketing boards or in 1947 with the Agriculture Act. Its end would then be marked by the first inroads made on the Common Agricultural Policy of the European Community by the imposition of quotas, the taking of land out of cultivation and other measures employed to reduce surpluses of food commodities in Europe. If the quantities of food produced are the measure, then the first increases in output of cereals, potatoes and sugar beet occurred in the late 1930s and the early years of the Second World War, but increases in the production of beef and other meats did not take place until the 1950s. If productivity as measured by yield per hectare is the criterion adopted, then the beginning can be estimated as the year of intersection of the negligible change in production before the Second World War with the positive relation between yield and time observed in the post-war period. This places the beginning between 1942 and 1947. On this criterion the revolution is not at an end: yield per hectare is still increasing, albeit at a reduced rate. Some criteria related to structural change show no discernible departures from a slow trend with the years. The number of farms has declined and the mean size of farms has increased fairly continuously since 1900; the decline in the landlord and tenant system has also been continuous since about 1910, as judged by the proportion of tenanted land. Farm labour has also diminished continuously for a long period, admittedly with increases during both the First and Second World Wars and accelerated declines thereafter. If technical advances are employed to date the revolution then there is no doubt that many of the signal discoveries were made long before they were applied on any scale in British farming.

All these considerations make it difficult to give a precise date for the beginning of the revolution or for its end – if indeed it has ended. What seems a sensible compromise is to state that considerable changes took place in British agriculture, beginning in the late 1930s and continuing until the mid- to late 1980s, and that these changes were of such magnitude that they constituted a revolution. The stagnation of 70 years was replaced by growth and agriculture was transformed. Combining compromise with convenience, the modern revolution is dated here from 1936 to 1986. Convenience stems from the fact that the time span of revolution is then a convenient fifty years and the end is a century after the date that has been taken to mark the end of the earlier revolution of the sixteenth century.

The economic and political background

Increases in agricultural output take place because farmers can discern an advantage in producing more. They consider the balance between the prices they obtain for the commodities they sell and the costs of producing them; they respond to this balance in the ways that can be expected of business people, increasing output when the balance is favourable, and vice versa. As discussed in the previous chapter, following the era of high farming of the last century, prices for farm produce were low because the British farmer had to compete with overseas suppliers for the home market. The major economic and political change that led to the revolution was the abandonment of the policy of free trade, which had provided the context for British agriculture for almost a century, and its replacement by protection from competition. It was as if, at long last, the country had heeded the words of Lord Melbourne who, speaking before the repeal of the Corn Laws in 1849, said 'To leave the whole agricultural interest without protection, I declare before God that I think it is the wildest and maddest scheme that has ever entered into the imagination of man to conceive'.

This reversal of policy began in a small way when the National government was elected in 1931. Ramsay MacDonald introduced the Wheat Act in 1932; this provided, through a levy on flour, a 'deficiency payment' to wheat growers. It was followed by the Cattle Industry (Emergency Provisions) Act of 1934. This provided a subsidy to farmers on the fat cattle they produced. In addition, duties were imposed on imports, including beef, veal, bacon and dairy produce. These controls on imports were reinforced by the Agricultural Marketing Act of 1933, which permitted marketing boards to restrict imports to protect the home market. It is doubtful whether these changes in agricultural policy were the result of deep-felt commitment on the part of the government to the idea that markets should not be left to international vagaries but controlled; it is more likely that they were a response to the obvious distress of the farming industry.

The Second World War led to far greater involvement of the government. Prices were closely controlled and in addition elements of compulsion entered the farming industry. Local committees were given extraordinary powers to ensure that the abilities of farmers and workers were harnessed to provide food from the land resource. The production of commodities that could be used as bulk foods rather than as delicacies was considerably increased. It was agreed when these measures were imposed that the controls made by the wartime coalition government should continue for a year when hostilities ceased. In fact, when the wartime legislation came to be dismantled, such was the realisation of the vulnerability of Britain's food supplies that it was replaced by legislation designed to protect and stabilise production by British agriculture. The Bill, introduced by Tom Williams, the far-seeing Minister of Agriculture in the newly elected Labour Government, passed through both Houses of Parliament without opposition. It became the now famous Agriculture Act of 1947, with provisions that dominated British agricultural policy until the United Kingdom entered the European Community.

The 1947 Act

The Act had the aims, set out in its opening provisions, of . . . 'promoting and maintaining, by the provision of guaranteed prices and assured markets . . . a stable and efficient agricultural industry capable of producing *such part of the nation's food and other agricultural produce as in the national interest it is desirable to produce in the United Kingdom*, and of producing it at minimum prices consistently with proper remuneration and living conditions for farmers and workers in agriculture and an adequate return on capital invested'. The Act provided for an annual review to determine the prices and assured markets for eleven commodities (the review commodities); these together accounted for about four fifths of the output of the industry. At its launch, prices were increased for most commodities by about 20% as an earnest of the government's intention. Although there was some ambiguity in the Act – as indicated by the phrase which has here been printed in italics – the Act gave stability to the industry through what was effectively an open-ended commitment. In the early years food for the consumer was rationed and the purchaser of food, both from abroad and from home sources, was the government through its Ministry of Food. As home production increased it became evident that some limits had to be set to the considerable exchequer support of the industry and the long-term commitment. The abolition of food rationing for the population in 1954 and the restoration of a market economy created problems. The 1947 Act was modified by the Act of 1957 so that guarantees applied only to an amount of produce that government thought fit to support. However, the principle of stability was retained. Reductions in commodity prices could only be made provided they did not amount to more than 9% in any three-year period. During the 1960s further changes were made. Rather than place most emphasis on price guarantees to provide support for the industry, the government gave greater emphasis to more indirect measures through grants and subsidies for specific purposes, many of which were designed to accelerate the adoption of new technology. The proportion of total exchequer support accounted for by these indirect means rose from about 25% of the total in 1955 to about 50% in 1972.

The principles of the 1947 and 1957 Acts continued to be adopted when the decision was taken to enter the European Community. Measures were taken to ease the transition to a different mode of agricultural support, bearing in mind that prices that farmers obtained for their produce in Europe were in excess of those that United Kingdom farmers obtained. In addition a policy was adopted to increase further home production of commodities through prices that were closer to those in Europe. Entry into Europe was, for most of the country's farmers, a return to the highly favourable and protected conditions of the early years following the 1947 Act. The Common Agricultural Policy, largely designed for a community in which farms were small and farmers numerous and poor, provided excellent returns for United Kingdom farmers who had larger and better capitalised farms and who, through experience and government support and encouragement, were well capable of adopting new and sophisticated technologies to increase production. Output increased throughout Europe; surpluses of commodities, exemplified by the so-called lakes of milk and wine and the mountains of butter, arose and it seemed that the complexity and ponderous nature of the Common Agricultural Policy was incapable of adjustment to prevent even further expansion of farm output. Eventually, agreement was reached between the member countries of the Community that some steps would have to be taken to reduce production and prevent the cost of support of agriculture taking the Community budget into bankruptcy. Changes in prices and quotas on production have been the result in recent years.

The cost of support

The economic measures taken to support United Kingdom agriculture since 1930 obviously reflect political decisions; looking back, each step seems to have been made in response to a present and immediate problem rather than to express some specific political philosophy. Perhaps governments always act on the basis of pragmatism! It was the realisation that farming was impoverished and that the rural economy was in decline that no doubt prompted the measures taken in the early 1930s to protect the industry. It was the advent of a war that was likely to be protracted that resulted in the measures required to augment home production at that time. After the war it was the shortages of food supplies throughout the world and the evident hunger of the British people that prompted further investment in home food production; it was probably not a measure of gratitude to the industry for its efforts during the war! In the 1950s the concern was with the immediacy of exchequer cost, whereas in the late 1960s and 1970s balance-of-payment problems were those of most importance. The import-saving role of home production of food, as expressed in the government's White Paper 'Food from our own Resources', then became paramount and its production was therefore encouraged.

The overall cost of the support of agriculture by government is not easy to estimate. When the Ministry of Food was sole purchaser of commodities it is difficult to state what the net cost to the state really was. Statistical data, collected on a systematic basis, are, however, available from 1955 onwards. These show that state financial support was £206 million in 1956–7 and had risen to £1481 million by 1986–7, a deceptively large increase which is, in fact, a reduction of about one quarter when inflationary changes in the value of money are taken into account. These figures include only the costs of market support and those of grants and subsidies given directly to farmers. They do not include the costs of research, advisory, educational and other services to the industry (Chapter 13). Richard Body in 1982 made calculations of the overall cost of agricultural support in the United Kingdom to conclude that the total cost of agricultural support was probably at least double the expenditure on market support and direct subsidy. He concluded that government support did not simply contribute to the farmers' incomes but exceeded the net income of the farming industry altogether. Long before the advent of the modern revolution there had, of course, been some support of the farming industry by government. Services had been provided to prevent the entry of plant and animal disease into the islands; control measures had been introduced on a considerable scale to prevent the depredations of pests and disease; educational material had been provided; and the government had funded innumerable investigations related to the state of the industry and its many facets. What has been unique and what has been the major determinant of the modern revolution is the unprecedented scale on which successive governments have financed the industry.

The change in land use during the modern revolution

The results of this investment in farming can be seen in several ways; the first is in the pattern of land use at the beginning and the end of the revolution. Table 4 summarises the changes that occurred in land use in the United Kingdom in the 50 year period from 1936 to 1986. The areas of both

Table 4. *Land use in the United Kingdom in 1936 and 1986*

	Area (thousand hectares)	
	1936	1986
Total area	24 085	24 085
Arable land	5 349	7 010
Permanent grassland	7 573	5 077
Rough grazings and deer forest	7 346[a]	6 461
Forest	1 405	2 328
Urban land	1 252	2 373[b]
Residual land[c]	1 160	936
Percentage of total land		
Arable land	22.2	29.1
Permanent grassland	31.4	21.1
Total farmed land	*53.7*	*50.2*
Rough grazings and deer forest	30.5	26.4
Total agricultural land	*84.2*	*76.6*
Forest	5.8	9.6
Urban land	5.2	9.8
Residual	4.8	3.9

[a] Corrected for area of deer forests, which were not grazed. These were excluded from the statistics of the time.
[b] Estimated from the population with the allowance of 47 ha per 1000 people.
[c] The residual component includes very small rural holdings, which are excluded from the statistics, together with mining areas and derelict land.

forested land and land devoted to urban use almost doubled; rough grazings diminished in their extent and so too did the sum of the areas of arable land and permanent grassland. About 54% of the land in the U.K. was devoted to agriculture in 1936 but by 1986 this had fallen to 50%. At the end of the period farming was practised on less land than at the beginning and it can be inferred that this land was probably of poorer value, since urban development usually takes place on the better land; the green-field sites preferred by developers are not found in remote areas of the Scottish Highlands.

The disposition of the land that was cropped or devoted to permanent pasture is given in Table 5. The values given are for the three years centred on 1936 and on 1986, respectively. They show that the area devoted to crops that can be gathered by combine harvesters – the cereals, pulses and oil seed rape – more than doubled while that cropped with root crops or fallowed (the traditional practices employed to ensure freedom of weeds in rotational farming) halved. Land devoted to fruit and vegetables and to temporary grassland hardly changed in area at all; permanent grassland was reduced by

Table 5. *The areas devoted to different crops in the United Kingdom in 1936 and 1986*

The values for 1936 are the mean values for 1935–7 and for 1986 the mean values for 1985–7.

	Area, thousand hectares		Increase
	1936	1986	1986 to 1936
Grain crops harvested with combine harvesters			
Wheat	745	1,994	+1249
Barley	360	1,904	+1544
Oats	989	110	−879
Mixed corn and rye	46	14	−32
Peas and beans	66	166	+100
Oil seed rape	—	288	+288
Total grain crops	*2,206*	*4,476*	*+2270*
Root crops and cleaning crops			
Potatoes	291	182	−109
Sugar beet	141	203	+62
Turnips and swedes	330	59	−271
Mangolds	97	12	−85
Bare fallow	162	43	−119
Total root crops	*1,021*	*499*	*−522*
Vegetables and fruit			
Vegetables	120	187	+67
Fruit	127	51	−76
Total vegetables and fruit	*247*	*238*	*−9*
Grassland			
Temporary grassland (leys)	1,732	1,737	+5
Permanent grassland	7,552	5,069	−2483

a third. These are massive changes in the disposition of land. Remarkable, too, is the increase of almost three million hectares in the area devoted to wheat and barley and the fall of almost one million hectares in that devoted to oats, associated in part with the decline in the horse population (see below). A hectare is one hundredth of a square kilometre; the change in acreage of these three cereals can be represented by a square with sides measuring 137 km!

Changes in the areas occupied by other crops were of smaller magnitude but in percentage terms very considerable. The area cropped with turnips and swedes fell by 82%, that with mangolds by 88% and the area under fruit by 59%. The new crop of oil seed rape came to prominence on entry into Europe and by the end of the period accounted for a greater acreage than that devoted

Table 6. *The numbers of people and animals in the United Kingdom in 1936 and 1986*

	Numbers (millions)		1986 as
	1936	1986	% of 1936
People			
Total human population	47.1	56.7	120
Farmers		0.291	
Farm workers	0.850	0.315	37
Animals			
Cattle	8.6	11.9	138
Sheep	28.0	29.9	107
Pigs	4.6	8.0	174
Poultry	80.6	116.6	145
Horses	1.0	—[a]	20

[a] The number of agricultural horses fell to fewer than 20 000 in 1966 and continued to decline so that by 1986 the negligibly small number of horses remaining consisted of museum animals of no significance to the farming economy.

to sugar beet, the crop that had been introduced and adopted in the period before and after the First World War.

Changes in numbers of livestock

Table 6 summarises the changes that occurred in the numbers of farm livestock during the fifty-year period. Cattle numbers increased by 37% but the greatest increase in the stocking of land was with pigs and poultry. Stocking is perhaps not the most appropriate term to use for these two species since they are almost entirely housed. They do, however, require land resources for the provision of their feed and the disposal of their excreta. Furthermore, the numbers of these two species at any one time do not reflect the numbers that are killed for their meat and are thus supported by the land. The sow is capable of producing two litters a year, and modern broiler chickens take only a few weeks to progress from a fertile egg to the now familiar product seen in the supermarkets. The number of pigs slaughtered each year rose from 6.2 to 15.6 million, that is by a factor of 2.5, while the number of poultry slaughtered increased from 47 to 532 million, an 11-fold increase. Sheep numbers did not increase greatly. Their number was reduced considerably by the severe winter of 1947 when about four million died of cold and starvation. It took several years for the sheep flocks of the country to recover from this disaster, but even so, the number of sheep and lambs slaughtered for meat increased during the period by 50%.

The horse deserves special mention. Agricultural horses had reached their peak in number in the first decade of the twentieth century and then, with the advent of the tractor, their number declined. Initially the decline was slow but it gathered momentum until, in 1958, it was regarded as no longer sensible to record their number in the annual census of agriculture. At the present time the horse can no longer be regarded as an agricultural animal. Farmland has, however, always supported large numbers of non-agricultural horses. In the early years of the century these were horses used for urban transport and numbered maximally about three-quarters of a million. These declined too but in recent years the number of horses kept for pleasure has increased considerably and the total number of horses kept on agricultural land is probably now of the order of 200 000. It should be remembered that the million agricultural horses were maintained virtually entirely on farm resources of feed. When they went they released almost half a million hectares of land, which could then be used for other purposes.

The increases in production of commodities during the revolution

Table 7 summarises the increases in the production of commodities from the farms of the United Kingdom during the 50 year period. The production of the major cereals wheat, barley and oats increased almost sixfold. This was due to a massive increase in the output of wheat and barley, for the output of oats fell to a quarter of that in the 1930s. The output of beans and peas also increased considerably and of course there was the new 'grain' crop of oil seed rape. Nearly 26 million tonnes of combinable crops were harvested in 1986 compared with only 4.5 million tons in 1936. Increases in the output of root crops for human consumption was less spectacular. This was because there was a restriction on output, imposed by the Potato Marketing Board and by the Sugarbeet Corporation respectively, designed to ensure that potatoes were not in excess of human demand and that sugar production at home did not interfere unduly with the preference given to imported sugar from the countries of the Caribbean. The area was restricted, too, by the need to ensure adequate intervals between these crops to control soil-borne pests and diseases. The quantity of root crops for livestock fell very considerably and so did the amount of hay harvested for them. In parallel was a considerable increase, not readily estimated with precision, in the amount of grass preserved as silage.

Meat production increased massively as well. Outstanding was the increase

Table 7. *The total quantities of commodities produced in the United Kingdom in 1936 and 1986*

	1936	1986	Ratio 1986/1936
Crops (thousand tonnes)			
Grain crops harvested by combine			
Wheat	1 503	13 910	9.3
Barley	744	10 010	13.5
Oats	2 013	505	0.25
Mixed corn	78	29	0.37
Rye	10	32	3.2
Beans	97	230	2.4
Peas	31	330	10.6
Oil seed rape	—	951	—
Total grain crops	4 476	25 997	5.8
Root crops for human consumption			
Potatoes	4 662	6 446	1.4
Sugar beet	3 503	8 120	2.3
Sugar derived from beet	530	1 323	2.5
Crops primarily for livestock			
Turnips and swedes	11 956	3 855	0.32
Mangolds and fodder beet	4 851	845	0.17
Hay	8 288	4 649	0.56
Silage	?5 000[a]	45 864	?9.2
Animal products (thousand tonnes)			
Meat			
Beef and veal	658	1 044	1.6
Mutton and lamb	197	291	1.5
Pork	191	754	3.9
Bacon and ham	230	206	0.9
Total meat from pigs	391	960	2.5
Poultry meat	80	934	11.7
Other animal products			
Wool	51	59	1.2
Butter	27	223	8.2
Cheese	56	257	4.6
Liquid milk (million litres)	4 587	7 453	1.6
Eggs (million dozen)	556	1 089	1.9

[a] No statistical information is available for the amount of silage produced in 1936. Most was produced on arable farms for the making of silage from grass was in its infancy. The figure given is acknowledged to be a guess.

Table 8. *Self-sufficiency in commodities in the United Kingdom and in the European Economic Community of ten nation states in 1986*

Values are the percentage of consumption that is met by production.

	United Kingdom[a]	Europe of 10
Wheat	118	132
Barley	150	129
Oats	97	
Oil seed rape	126	
Potatoes	88	103
Sugar (refined)	56	135
Fruit (other than citrus)	22	83
Vegetables	65	101
Beef and veal	96	108
Mutton and lamb	84	76
Pork	102 ⎱	102
Bacon and ham	45 ⎰	
Butter	81	133
Cheese	65	107
Eggs	98	102
Whole milk powder	396	334
Skim milk powder	215	118

[a] The data for the United Kingdom are the mean of three years 1985–7.

in the production of poultry meat, which rose almost twelvefold. By contrast, the overall increase in the production of meat from cattle, sheep and pigs was by only 85%. The amount of milk sold for liquid production also increased. The amount actually produced was considerably greater than the liquid sales suggest; the considerable excess over liquid demand was used to manufacture butter and cheese and was also dried to make a variety of milk powders. Twice as many eggs were produced at the end of the fifty-year period than at its beginning. Even wool production showed an increase greater than that which might be inferred from the size of the nation's ewe flock.

The most striking aspect of the revolution was that the United Kingdom was transformed from a grain-importing to a grain-exporting country. In the 1930s our imports of wheat and barley were about six million tonnes a year: about three times what we produced. In 1986 our net *export* of these two cereals was also six million tonnes, although bread wheat continued to be imported at this time. Despite the considerable amounts of grain which were used to support the animal industries – accounting for close on ten million tonnes – the United Kingdom produced considerably more than was needed to meet home demand. As Table 8 shows, for many other commodities the output

Table 9. *Yields of crops per hectare in 1936 and 1986, and other indices of productivity*

The values for 1936 are the mean values for 1935–7 and for 1986 the mean values for 1985–7.

	1936	1986	Ratio 1986/1936
Crop yields ($t\ ha^{-1}$)			
Wheat	2.16	6.44	2.98
Barley	2.00	5.06	2.53
Oats	2.02	4.70	2.33
Beans	1.94	3.80	1.96
Peas	1.79	3.60	2.01
Oil seed rape	—	3.18	—
Potatoes	16.5	36.5	2.21
Sugar beet	22.7	39.6	1.74
Turnips and swedes	33.7	65.3	1.94
Mangolds	46.2	74.3	1.61
Other indices of productivity			
Milk yield per cow (litres/year)	2400	4900	1.79
Eggs per bird (number/year)	149	259	1.79

of United Kingdom agriculture was close to that required for self-sufficiency and in the European Community of ten countries most commodities were produced in excess of the Community's needs. Taking the more parochial view and considering the United Kingdom alone, of all the food that could be produced in our temperate climate, about 80% was grown on our own farms. This contrasts with the 1930s when the country's farms produced only about 30% of the temperate food its people required. Furthermore, and emphasising still more the magnitude of the accomplishment, the population of the United Kingdom had increased by almost 20% in the fifty-year interval; there were more people to be fed and more food was produced for each one of them.

The improvement in productivity

The increase in the output of commodities from a diminished area of land must mean that the efficiency of the farming industry in using the land resource had considerably improved. Such increases in productivity are more easily assessed for crops than they are for livestock; one can simply compare yields per hectare in 1936 and 1986 to obtain a good measure of progress. Yields vary with the weather and the data given in Table 9 are the means of three years centred on the beginning and end of the fifty-year period. Cereal yields doubled or nearly trebled, and the increases in the yields of other crops were also about doubled. These increases can be compared with those of the previous fifty years, which were shown in Table 3. Then, improvement in productivity was measured as a meagre few percentage units rather than in terms of doubling or trebling.

There are no similar statistics to enable calculation of the changes in output of animal products per unit of land. Some data are available, however, which measure the increase in the productivity of the average animal. Thus annual milk production per cow increased by 65% and annual egg yield per bird by 79%. There is no doubt that more cows were kept on smaller areas of grassland and so the increase in production per hectare must have been considerably greater than the milk yield figures alone would suggest. The Cambridge agricultural economists have for many years summarised the results of surveys of the performance of pig farms. Their series does not go back to the 1930s but shows that the average number of piglets raised by a sow in a year rose from 10.6 in 1946 to 20.3 in 1986 (see also Chapter 12). At the same time the number of kilograms of feed used to produce one kilogram of gain in body mass fell from 4.95 to 2.88. Admittedly this example of the increase in the efficiency of producing meat is drawn from a selected population but is indicative of the technical improvement that took place.

The ten-yearly censuses show that in 1931 the working population of the United Kingdom was 16.3 million and those directly engaged in agriculture – that is, farmers and farm workers – numbered 1.3 million or 8% of the total. In 1961 the working population had increased to 26.7 million and those engaged in agriculture had fallen to 0.63 million or 2.4% of the work force. The figures for the number of farmers and farm workers, given for 1936 and 1986 in Table 6, although perhaps less accurate than the census figures, show much the same massive reduction in the numbers working on the land. Obviously, since production had increased very considerably, output per person had increased even more. Again, the adoption of new technology enabled this to be accomplished.

An economic appraisal

When discussing how to measure the aggregate output and the overall efficiency of farming, it was pointed out (Chapter 1) that one way was to express output and input in monetary terms. This is feasible provided that the value of money does not change. In dealing with the modern agricultural revolution the value of the pound sterling was about 14 times greater at its beginning than it was at its end. This complicates matters. Marks and Britton, using a general price index of goods and services, have adjusted the estimates

Table 10. *The output value, input costs and farming income from 1935-7 to 1985-7, calculated in terms of 1986 money values (billions of pounds, £10⁹)*

The data are those calculated by Marks and Britton (1989) from the returns made by the agricultural departments.

Average of years	Gross output	Gross input	Gross product	Net product	Wages, etc.	Farming income
1935–7, 1938–9	7.04	3.00	4.03	3.81	2.34	1.47
1945–7	8.54	2.70	5.84	5.47	2.94	2.53
1955–7	12.23	6.08	6.16	5.42	2.77	2.65
1965–7	13.04	6.76	6.28	5.39	2.70	2.68
1975–7	14.67	7.72	6.95	5.45	2.56	2.89
1985–7	12.15	6.50	5.64	4.20	2.90	1.30

of the financial returns of the country's farms made each year by the agricultural departments, to the common basis of 1986 money values. They employed a general index of prices to do so; their results are given in Table 10. Some explanation of the headings is necessary. Gross output is the total receipts obtained by farming. It includes produce sold and, in addition, the sums provided by the government in compensatory payments and production grants. Gross input is the sum of all the costs which the farmer incurs save those on hired labour, interest on borrowed money, and rent. These input costs can be regarded as the physical inputs into the farming system and comprise seed, fertilisers, feeding-stuffs, machines and their maintenance. The difference between gross output and gross input is the gross product. Net product is this sum less depreciation on machines, vehicles and buildings and it is this net sum that has to meet the cost of labour, interest on borrowing and rent before it leaves a profit to the farmer.

The results of this complicated accountancy show that gross output about doubled during the revolution; this is in broad accord with the data, given earlier, on the tonnages of crops and livestock produced. The costs of inputs, however, rose by more than twofold, and in consequence both the gross product and the net product were not much greater at the end of the period than at the beginning. When labour costs, rent and interest on borrowed money are deducted, farming income in 1986, expressed in real terms, was much the same as, or even less than, it had been before the war. If a revolution means that things turn full circle, eventually to restore a *status quo*, then there is little doubt that in economic terms the farming industry passed through one in the course of the fifty years ending in 1986.

Some limitations of estimates of productivity

Although there is no doubt that the modern revolution resulted in considerable increases in yield per unit area of land and in output of commodities per person in the agricultural work force, the economic consideration above raises questions about whether these are sufficient measures of the efficiency of the industry. The financial data in Table 10 show that increased output was only obtained as a result of increased inputs. These inputs were certainly not the land area or the agricultural work force, for they both diminished. Increased yields of crops per hectare were associated with an increased use of fertilisers and agrochemicals; the reduction in the manpower needed for cultivation of the land and for the care of livestock was associated with mechanisation of all agricultural processes. Fertilisers, lime, herbicides, pesticides, protein-rich feeding-stuffs, tractors, implements, diesel fuel, electrical power, machine maintenance and specialised buildings are all resources that are purchased by farmers and are not intrinsic to the farm. Their use increased on an unprecedented scale during the revolution, for they were integral components of the technological changes which the farming industry undertook. Some cognisance has to be taken of the fact that, far from being an isolated sector of the economy as it was in the eighteenth century, dependent only on its own resources of land and labour, farming by the 1980s had become in modern jargon, 'non-sustainable', that is dependent on the conventional manufacturing industries for additional and essential resources.

Questions can thus be asked about how far the resources that are 'upstream' of the farm should be considered in assessing the efficiency of the industry. One can argue that to the farm labour force should be added the workers in the fertiliser industry, the feeding-stuffs industry, the engineers designing and building tractors and implements, and the small firms which undertake repairs to them. An estimate made in 1974 of the upstream labour force showed that there were 60 000 people engaged in tractor manufacture, 29 000 in feeding-stuffs manufacture, 12 000 employed in making fertilisers and 23 000 concerned with agricultural education and advice to the industry. The numbers in these groups alone are equivalent to about half the number of farmers in the country. To these must be added implement manufacturers, others concerned with transport to and from the farm, those who provide specialist services – such as veterinary care, machinery repair and financial help – and the large number of representatives of firms who call at the farmer's door. The work force of 600 000 in 1986 may well have been supported by an equal number in agriculturally oriented activities upstream or contiguous to the industry. Extolling the virtues of the farming industry, on the basis of increased production per person, needs to be tempered somewhat in recognition of the additional manpower employed outside the farm in maintaining its productivity.

In an analogous way, one can argue that the justifiable pride taken in the increase in crop yields per hectare should be modified since the land was not the only resource deployed; other industries were involved in providing the tools that farmers used to such good avail in increasing productivity. Indeed, looked at in this way a characteristic of the modern revolution has been the extent of its support by other sectors of the economy.

The industrial support of farming

Another way in which to consider the extent of support of the farming industry by conventional industries is to express the inputs not in monetary but in energy terms, that is in terms of the energy, mostly derived from coal, oil and natural gas, which is needed to fabricate the products that the farmers buy. This has the advantage that these 'support energy' costs are free of the inflationary changes affecting the value of money. The approach was first suggested by Howard Odum, an American ecologist, who was primarily interested in the operation of ecosystems and the way in which the energy, trapped by green plants, flowed through them from species group to species group. A series of conventions has arisen in applying these energy accounting methods to human endeavours and these are outlined in most accounts of them. In Britain, in our fifty-year period, the overall increase in the annual use of support energy was largely accounted for by fertilisers and fuel and by 1986 the agricultural sector accounted for almost 4% of the country's expenditure of energy, most of which came from fossil sources. What is remarkable is that when the whole of the food system is considered, including the manufacturing sector, wholesale and retail distribution and home storage and cooking, the energy cost of providing food on the consumer's plate is ten joules for every joule of sustenance. The revolution between the 1930s and 1980s was thus not limited to the agricultural component of the whole food provision system: it affected equally all other sectors.

As far as farming is concerned the modern revolution can be regarded as a period in which the power base of the industry changed. This statement could be simply taken to mean the replacement of the horse by the tractor, a process analogous to previous revolutions in which the horse replaced the ox as the motive power unit or in which the ox replaced the human for the onerous tasks of tilling the soil. The concept is, however, much broader and has greater implications. Firstly, the move from horse to tractor substituted a biological and sustainable energy source for one based on extractive industry. Secondly, replacement of the motive unit was not the only change. Reliance on biological nitrogen fixation by clovers, and emphasis placed on the value of animal excreta for fertility maintenance and on rotational farming for the control of weeds, pests and diseases, were replaced by continual annual inputs of artificial fertilisers and agrochemicals. The energy required every year to synthesise these materials constituted a major power input to supplement the annual power input of solar radiation on which farming had hitherto solely depended. Emphasis is to be placed on the word 'supplemented'. It is highly improbable that the present output and the present yields of crops could be obtained without the use of these modern inputs. They thus supplement the biological resource of solar radiation.

The new technologies

What has been remarkable about the modern revolution in agriculture, distinguishing it from earlier ones. is that the technologies that have been adopted are for the most part based on the application of a mature agricultural science. In many instances the impact of this science has been direct. A new, improved variety of crop plant or an effective vaccine for combating a disease in animals can obviously be applied directly on farms provided there has been adequate test of the variety's suitability or the vaccine's efficacy. With many new discoveries and ideas generated under laboratory conditions or on experimental farms, however, application to commercial circumstances inevitably entails modification to meet particular conditions and a rigorous economic appraisal before acceptance. Much of this developmental and appraisal work was done by leading farmers but throughout the period the agricultural departments placed considerable emphasis on such activities. The National Agricultural Advisory Service of the Ministry of Agriculture was superseded by a new body, the Agricultural Development and Advisory Service (A.D.A.S.), which had its own farms – the Experimental Husbandry Farms – on which to try out new methods under more applied conditions. In Scotland the advisory and developmental work was organised so that there was a direct link with the research and educational activities of the three Scottish Colleges of Agriculture and University departments of agriculture. The extent to which the government supported education in the science of farming, in agriculturally oriented scientific research, in the development of research findings and in advising farmers on how to adapt them to the conditions peculiar to their own farms, can be judged by the number of people employed by the state in these activities. In the 1970s, at the height of the revolution, there were 6484 people employed in Britain in the advisory and development services, 6736 in the agricultural research service and 1300 senior people concerned with agricultural education in the universities and colleges. The total number deployed to fuel the revolution amounted to one for every 15 farmers or one to every 700 hectares. It is understandable that the pace was considerable.

The pace of change had, in addition, an internal momentum. As new technologies were adopted by the forward-looking and innovative farmers, with resultant increases in production and hence of income, so others, looking over their fences, saw what was being achieved and emulated them. As a result overall production in the country increased and prices in real terms tended to decline. To preserve income, the progressive farmers looked for yet new ways in which they could increase production and, if possible, simultaneously reduce their costs. They were copied; production rose still further to result in lowered real prices and thus promote a further cycle of adoption of new technology. This process in which increased production leads to the ready acceptance of new methods and techniques has been called 'the technological treadmill'. The term implies a servitude, but most farmers did not regard it as such. Farmers take pride in the accomplishment of higher yields and in other indices of good husbandry, as any visit to an agricultural show plainly demonstrates. Nevertheless the treadmill effect certainly contributed to the avidity with which farmers embarked on technological change.

In the chapters that follow, some of the scientific discoveries, which were then developed to enable the revolution, are described, giving a prominence to those men and women of genius who were largely responsible. This necessarily singles out a few from the many thousands of scientists who were also involved, both in the United Kingdom and throughout the world. Their intellectual contribution to the revolution was immense; although the massive increase in output from the land can be regarded as the farmers' response to a favourable economic climate, without the scientific base and the nascent technology this would not have been possible.

Selected references and further reading

Beresford, T. (1975). *We Plough the Fields : Agriculture in Britain Today*. Penguin, Harmondsworth, U.K.

Body, R. (1982). *Agriculture: The Triumph and the Shame*. Temple Smith, London.

Britton, D. K. (ed.) (1990). *Agriculture in Britain: Changing Pressures and Policies*. C.A.B. International, Wallingford, U.K.

Harris, S., Swinbank, A. and Wilkinson, G. (1983). *The Food and Farm Policies of the European Community*. John Wiley and Sons, Chichester, U.K.

Henderson, W.M. (1981). British agricultural research and the Agricultural Research Council. In *Agricultural Research 1931-1981*, ed. G.W. Cooke, pp. 3–113. Agricultural Research Council, London, U.K.

Leach, G. (1976). *Energy and Food Production*. Science and Technology Press, Guildford, U.K.

Marks, H.P. (1989). *A Hundred Years of British Farming: A Statistical Survey*, ed. D.K. Britton. Taylor and Francis, London.

Ministry of Agriculture, Fishery and Food (1975). *Food from Our Own Resources.* Command 6020. Her Majesty's Stationery Office, London.

Ministry of Agriculture, Fishery and Food (1979). *Farming and the Nation.* Command 7458. Her Majesty's Stationery Office, London.

Odum, H.T. (1971). *Environment, Power and Society.* John Wiley and Sons, London.

Tracy, M. (1964). *Agriculture in Western Europe.* Jonathan Cape, London.

PART TWO
The science and technology of the modern agricultural revolution

Agricultural science is not a single discipline, save that the men and women who are involved have a common purpose in improving the provision of food from resources of land and climate. It draws heavily on the basic physical and biological sciences and, indeed, contributes to them. Much of agricultural science is engendered by realisation of real problems in agriculture, some by an appreciation of likely future ones and some by curiosity and wonder about the nature of the living things on which we all depend. Each year literally millions of scientific papers are published which advance the front of our knowledge about the many factors which contribute to the productivity of soils, plants and animals. To summarise such a wealth of information and give recognition to the many who have made contributions to our understanding would be a Herculean, if not an impossible, task. In the chapters which follow there has, of necessity, been selection in which an attempt has been made to give prominence to those discoveries and inventions that have had the greatest impact. Furthermore, science is a vast continuum and some of the signal scientific advances were made long before they were applied to what was in the 1930s a craft industry to transform farming into the science-based agricultural industry of the present. These early discoveries are mentioned where appropriate.

3 Problems of measurement

Over the dais in the great lecture hall of the Chemisches Unterrichtslaboratorium of the ancient University of Leipzig is the phrase 'Gott hat alles nach zahl, masse und gewicht geordnet' (God has ordered everything according to number, size and weight). This quotation was well appreciated by Justus von Liebig, the first of the great agricultural chemists, and he quoted it. The importance of quantitative measurement has indeed always been appreciated by scientists and is at the heart of the many disciplines that make up agricultural science. This is not to denigrate studies, the results of which can only be expressed in qualitative terms, or to minimise the importance of purely descriptive work; after all, discoveries of the life cycles of a parasite, or of the agent which causes a plant disease, are major scientific accomplishments. However, much of agricultural science is concerned with exploration of the many processes that might result in higher or more efficient growth and productivity of plants and animals. This necessarily entails measurement and the use of experimental method. The new technologies that enabled farmers to exploit favourably economic circumstances were dependent on the discovery of new ways in which to conduct and evaluate experiments. Furthermore, these discoveries, which facilitated agricultural productivity, have had considerable repercussions in other branches of science. Much of modern statistical technique and methodology, applied to such diverse subjects as quality control in manufacturing industry or assessment of the pathogenicity of microorganisms or the conduct of social surveys, originated in work carried out with improvement of agriculture in view. The development of the new statistical and experimental methods arose from the vexed problems of how to measure the effects of different treatments on crops. An account of the difficulties that were encountered before solutions were reached is desirable to provide the context in which the discoveries were made.

Experiments with crops

In the eighteenth century and earlier the term 'experiment' was employed in what is now an archaic sense to mean a trying out. A new crop was grown, a new fertiliser used or some new cultivation applied on one large area and the outcome noted. There was no element of comparison, except that subsequent adoption of a process entailed some subjective comparison with what went before. Some investigations were made, however, in which there were direct comparisons of one procedure with another, but these were few. Most consisted of large plots, one of which was a 'control' and received no experimental treatment. The limitations of this approach were realised quite early. In 1849 J.F.W. Johnston published a book which recommended that 'everything should be done by weight and measure' and that 'two experiments of the same kind, one to check the other, should always be made'. He made the further point that 'Two portions of the same field – supposed to be equal in quality and for that reason selected as especially fitted for experimental trials – may naturally yield very considerable differences in crop, even when no manure is added to them'. A further difficulty is, of course, the weather; no two seasons are identical.

The classic experiments with wheat at Rothamsted were commenced in 1843 and consisted of single plots arranged side by side on the same field and manured in different ways. The experiment was continued year after year, giving a replication over time; Lawes and Gilbert only published the results after a lapse of 20 years. They recognised that the results they obtained were likely to apply only to the soil at Rothamsted and they therefore repeated the experiment at other centres: Holkham in Norfolk, Rodmersham in Kent and later at Woburn in Bedfordshire. The scale of experimentation at Rothamsted at this time was considerable; the experiment with turnips, also commenced in 1843, consisted of 24 separate treatments, each a long strip down the length of the field and occupying a third of an acre. In 1845 the 24 main treatments which ran the length of the field were crossed with four additional ones to give 96 separate unreplicated plots. This gave a more compact layout and was the forerunner of the chessboard layout of plots. This layout was used in the way that Lawes and Gilbert had used it to examine how different varieties of a crop responded to different levels of manuring; strips of each variety were crossed with the manurial treatments. It was also used by E.S. Beaven to examine individual genetic isolates by growing seed from particular crosses in small plots, each one metre square. Beaven replicated these chessboards 20 times so as to obviate differences in yield that could arise from differences in soil fertility.

Experimenters were only too aware that variations in fertility within a field could affect their results, and considerable ingenuity was employed in designing arrangements of plots within a field which were 'balanced', that is, were so disposed that plots given any one treatment occurred on the north of the experimental area as well as the south and in the east as well as the west. One such arrangement, in an experiment in which the plots were rows of crops, was to group the plots around a central one such that the distances of the rows representing one treatment, when added up, were the same on both sides of this central plot. Another, devised by Beaven in 1920, was the half-drill strip design for comparing two varieties. This design remained in vogue for several decades, largely because of its convenience. One half of the seed drill was filled with seed of one variety and the other with seed from a different variety. As the drill moved up and down the field, turning at each headland, the order of the rows of crop were A B B A A B B . . . etc. It was thought that this would obviate effects of fertility gradients across the line of the rows because every plot of variety A was flanked by one of B and one of A. Had the sequence been A B A B A B . . . etc., a fertility gradient across the rows would have favoured one or other of the varieties.

The advent of statistical methods

The theory of probability and the measurement of variation associated with observations had been well developed from its early beginnings in the eighteenth and early nineteenth century. Ways of estimating the error attached to means of large numbers of observations had been worked out, and it was the hope that these sorts of approach, based on statistical theory, could help in the analysis of the results of agricultural experiments. What was needed was the computation of estimates of the errors attached to the means obtained from field trials and some rational way in which conclusions could be drawn about whether one treatment was really better than another. The basic difficulty in arriving at a solution was that the number of replications in field experiments was usually small and far removed from the very large numbers which the statisticians of the time employed in their calculations. A start was made in a collaboration between an agriculturalist, T.B. Wood of the Cambridge School of Agriculture, and a mathematical astronomer, F.J.M. Stratton. They estimated from large numbers of observations on the weights of mangolds that the 'probable error' attached to single plots of a size greater than one fortieth an acre was about 5% of the produce of the plot, suggesting to them that the error of differences between adjacent large plots would be about 7%. In 1911 W.B. Mercer and Sir Daniel Hall reported a trial designed to measure variability in the wheat crop. They grew one variety on an acre of a very even field, divided the acre into 500 separate plots, harvested each one and weighed the grain and straw produced. They showed that the variation encountered was that to be expected from theory; the 500 yields were distributed evenly about the mean, conforming with the classical Gaussian or normal distribution. The importance of this paper lies not so much in the data they presented as in the appendix, which was written by 'Student' (the pseudonym of W.S. Gosset). Although he had published the salient information in an earlier paper in 1908, Gosset brought to the attention of agronomists the finding that a test of the significance of differences between the means obtained in an experiment could be made, even though the number of observations was small. The test which he devised was called the '*t*-test'.

Gosset had been educated at Winchester and at New College, Oxford, where he had read chemistry and mathematics. On graduation he entered the Guinness brewery in Dublin as a chemist. There he developed close links with Beaven, who was a breeder of malting barleys. The collaboration arose through a mutual interest in studies on the breeding of better barleys being carried out by the Department of Agriculture in Ireland. Gosset's major contribution was that he solved the problem of how to estimate the significance of a difference between the mean results obtained when two varieties – or indeed any other two imposed factors – were compared in experiments which were not massively replicated.

R.A. Fisher and modern, statistical methods

In 1919 Sir John Russell, Director of Rothamsted, appointed Ronald Fisher to re-examine the accumulated results of the long-term trials which had been commenced by Lawes and continued by others. Fisher stayed at Rothamsted for 14 years before moving to University College, London, and the Galton chair. During that relatively short period he laid the foundation for the whole of the modern fabric of mathematical statistics. Fisher had been educated at Harrow and at Gonville & Caius College, Cambridge, and had developed an interest in both statistics and genetics while reading for his degree in physics and mathematics. After graduation he spent a further year at Cambridge studying statistical theory with F.M.J. Stratton who had undertaken the earlier experiments with Wood. Whether this link with experimentation in farming made him decide to accept the job at Rothamsted – rather than an attractive one at the Galton laboratory – we will never know. At Rothamsted he had freedom to pursue both interests. He did some work on the long-term experiments, as Sir John had requested, but his main contributions came from the development of new concepts and new ways in which to design experiments, estimate parameters from them and test their significance. Quite early, Gosset approached him about his *t* distribution, for Gosset had been unable to generalise it. This Fisher did, to devise the *z* distribution. He systematised the analysis of numerical data by devising the analysis of variance and showed that by insisting on the random allocation of plots within the confines of an experimental design the problem of the independence of errors could be overcome. He summarised his earlier contributions in his book *Statistical Methods for Research Workers* (1925–34). This went through 14 editions and was translated into French, German, Italian, Russian, Japanese and Spanish. It had immense influence on agricultural experimentation. Fisher devised methods of factorial analysis so that combinations of different factors affecting yield could be studied simultaneously and applied these factorial designs in experiments to measure effects of different levels of fertilisers. He studied latin square layouts and randomised block designs, and devised variants of them. He devised the principle of confounding in experimental design which effected considerable economy because it made possible the isolation of main effects of treatments and their important joint effects by deliberately incorporating their higher-order interactions into the error term. He developed ways in which concomitant variation could be taken into account through analysis of covariance, and he rationalised the design and analysis of experiments on orchard and other perennial crops. There were few aspects of agricultural science which escaped his acute mathematical insight, from the estimation of bacterial numbers to the analysis of particle size in soils by examination of their rate of sedimentation from suspension. In their account of Fisher's life, Frank Yates and Kenneth Mather, who both worked with him, wrote: 'In spite of the early controversies, the new ideas on experimental design and analysis soon came to be accepted by practical research workers, and the methods have now been almost universally adopted, not only in agriculture but in all subjects which require investigation of highly variable material. The recent spectacular advances in agricultural production in many parts of the world owe much to their consistent use'.

Subsequent developments

These new statistical methods, initiated by Fisher, were rapidly developed and applied to aspects of agriculture other than crop husbandry. David Finney developed methods for biological assay: the procedure where responses of animals to vitamins or hormones could be employed to estimate the quantities of the substance in a diet or preparation. Jay Lush in the United States applied them to problems related to the genetic improvement of livestock and, in continuation of Fisher's interest in genetics, quantitative methods were devised by K. Mather and others to deal with inheritance of these characteristics in plants. Agricultural courses in universities and colleges were modified to include study of the new methods, and research centres throughout the world emulated Rothamsted in appointing statisticians to their staffs to provide guidance to experimenters on the design of their experiments and the interpretation of the results obtained. Nor were the aspects of experimental design neglected. Frank Yates and, later, Rosemary Bailey made considerable strides in the devising of designs that were rigorous in terms of the validity of the conclusions that could be drawn from them.

Statistical analysis involves a large amount of tedious arithmetic. Initially this was done by hand and later by mechanical calculating machines. With the advent of electronic computers in the early 1950s the formidable work involved in statistical computation was immediately reduced. In 1950, Rothamsted, before the purchase of its first computer, analysed annually the results of 500 experiments; after the introduction of computers the numbers rose in 1964 to 3 400. Thereafter they continued to rise but because of different methods of recording the activities and the fact that ever more complex analyses became possible, comparable figures are not available (Gower, 1985). At the same time it was becoming obvious that experiments with the same design could be analysed using the same computer program and Frank Yates and his Rothamsted colleagues then developed a large general statistical package of computer programs named GENSTAT, which catered for nearly every type of statistical programme likely to be encountered by experimenters. GENSTAT was immediately highly successful and is used throughout the world by agriculturists and other scientists alike.

Practical application

The very large number of experiments that must have been carried out during the modern agricultural revolution is evident from the discussion above. Questions arise about how the vast amount of information obtained could have been summarised into relatively simple recommendations to farmers about optimal courses of action. This problem was considered by statisticians quite early in the period. E.M. Crowther and F. Yates inveigled their colleagues into the onerous task of abstracting information from all the fertiliser trials that had been conducted in Britain and Europe since 1900. Five thousand were identified and analysed statistically. This resulted in the formulation of response curves for the agriculturally important crops and from these the optimum amount of fertiliser was identified. The point where significant increases in economic yield were no longer demonstrable was taken as the rather arbitrary point for defining the optimum fertiliser application. Crowther and Yates found that on most soils nitrogen was required. Where the land had been treated with farmyard manure the responses to added phosphorus and potassium fertiliser on particular soils might be small and savings on the wartime imported phosphate fertilisers could be made. A knowledge of local conditions was necessary for the sensible application of these fertilisers. Where fertiliser was expected to improve yield the optimum amounts recommended were 1.2 cwt of nitrogen, 0.5 cwt of phosphate expressed as P_2O_5, and 0.5 cwt of potash, expressed as K_2O, per acre in normal agricultural soil (1 cwt $\approx$ 50.8 kg). These results, published in 1941, formed the basis of fertiliser practices recommended to farmers during the Second World War, for the rationing of fertiliser supplies, and for import policies adopted during the blockade of normal trade. Great as their value was in rationalising, from a vast array of information, the use of fertilisers in time of war, Crowther and Yates in their analysis also reached the important conclusion 'that the current level of nitrogenous manuring, both absolutely and relative to that of other fertilisers is too low and considerable increases in production would follow greater use of nitrogen'. This conclusion undoubtedly helped later in the promotion of the use of nitrogenous fertilisers to enhance crop yield.

With the adoption of sound experimental design and efficient methods of statistical analysis of the results obtained from them, a return could be made to the approach made by Lawes when he replicated his classic wheat experiment on different farms. Coordinated experiments were organised in collaboration with the agricultural advisory services to examine practical problems as they arose. Some of the designs of these were sophisticated, and they yielded results that were immediately applicable and relevant to different soil types. An analogous development was undertaken by the National Institute of Agricultural Botany to test, under different field conditions, the new crop varieties that were emerging so as to arrive at practical recommendations. With animals, Raphael Braude at the National Institute for Research in Dairying commenced a series of coordinated experiments with bacon pigs to examine the validity of nutritional findings and the extent of variation from centre to centre under a variety of husbandry conditions; Frank Elsley at the Rowett Research Institute commenced a similar series with breeding sows. All these owed much to the development of statistical method and the appreciation of the importance of experimental design.

Farm economics and management

Farmers, however, needed something more than facts about whether one particular procedure was better than another in terms of crop yield or animal production. They were concerned with financial return and with the overall economics of their farms. Before the modern revolution, farm economists were not really concerned with the economics of innovatory change. Rather, most of their work was concerned with descriptive and retrospective cost accounting. Such studies sometimes led to identification of an attribute of the management of a farm that if altered could lead to enhanced profit, but the main use of cost accounting was in the assessment of the profitability of particular sectors of the industry. Nevertheless, these early studies did result in what has become a standard way in which farm enterprises can be costed and in which alternative courses of action by the farmer can be assessed. The first steps in this direction were taken in 1927 by J.S. King of the Department of Agricultural Economics of Reading University, when he distinguished 'overhead costs' from 'prime costs' in the analysis of farm accounts. Using these terms, he separated the costs that were fixed by the size of the farm, its rent, overheads and labour force, from those of the inputs into particular enterprises, which could vary appreciably. Gross profit he defined as the difference between the returns from the enterprise and the prime cost; this gross profit had then to be set against the overhead costs. This schematic way of using farm accounts is now called 'gross margin analysis'; prime costs are called 'variable costs', and overhead costs 'fixed costs'. Developed by V. Liversage in Northern Ireland and by D.B. Wallace at Cambridge, gross margin analysis proved to be a simple and effective way of budgeting the feasibility of a change in policy on an individual farm in terms of its financial outcome. Its application has been much aided by the annual publication by Professor J. Nix of a handbook that provides up-to-date costs of components of production systems.

In the 1920s attempts were made to generalise the economic information obtained from the accounts of farms. Statistical methods were employed, notably by Mordecai Ezekiel and J.D. Black in the United States. They were concerned in finding ways in which inputs of the various factors of production could be optimised and used to increase financial returns of individual farms. The approach was entirely empirical: there was no underlying theoretical structure. In 1949 Earl O. Heady proposed that such approaches should accord with accepted economic theory; his approach constituted a real advance. Heady spent the whole of his life, from his student days until retirement, at the University of Iowa; he was responsible for a resurgence of interest throughout the world in quantitative and predictive methods in agricultural economics. His book *'Economics of Agricultural Production and Resource Use'* was seminal. There he developed ideas that had emanated from classic economic theory of the nineteenth century, particularly the theory of marginal analysis (propounded by Alfred Marshall) and the theory of the firm.

Econometrics and Earl O. Heady

The new subject was called 'farm econometrics' and its starting point was the mathematical formulation of what were termed 'input–output relationships' for each of the production enterprises to be found on farms. Emphasis was given to the continuous nature of the responses of saleable products to many and varied inputs and the inter-relationships between these different inputs. The concept arose of multidimensional response surfaces determined by many factors. Each surface was then analysed in the light of price and cost structures, uncertainty about these prices and costs and the availability of capital. The analysis provided solutions that showed the optimal disposition of resources for any particular set of financial circumstances. The economists avowed that these approaches reflected the reality of the complexity of the farm business and the nature of the decisions that farmers were called upon to take. Heady's approach was taken up avidly by others concerned with the problems facing the farmer in making decisions, not so much about whether to use a new piece of technology, but in deciding how much of that technology to apply. J.H. Dillon at the University of New England, Australia, developed ways of conducting experiments to apply the principles involved, drawing heavily on the contemporary work in experimental design, while J.B. Dent showed how they could be extended to animal production enterprises.

It was perhaps inevitable that Heady's approach annoyed both conventional farm economists – who thought he was moving beyond the bounds of the data available to him – and the biological and physical scientists, who, concerned with increasing agricultural production with little immediate thought given to the economic usefulness of their findings, believed the work on responses an unnecessary duplication of their own. This latter attitude was no doubt fostered in the United Kingdom by the fact that the Agricultural Research Council did not undertake research in farm economics; this was the prerogative of the Ministry of Agriculture. It is certainly true that the experimental agriculturalists had long been aware of the complexity of the determinants of yield of both crops and stock. Fisher had written in 1926, 'No aphorism is more frequently repeated in connection with field trials than that we must ask Nature few questions, or ideally, one question at a time. The writer is convinced that this view is wholly mistaken. Nature he suggests will best respond to a logically and carefully thought out questionnaire, indeed if we ask her a single question she will often refuse to answer until some other question has been discussed'. Furthermore, the many experiments undertaken to examine effects of experimental treatments singly and in combination had certainly given rise to the concept of a response surface years before. Other scientists, concerned with understanding the biochemical and physiological processes underlying the responses of crops and animals, thought that Heady's largely empirical approach to them was rather naive. The essentially economic arguments that derived from the explosive growth of mathematical econometrics at that time were not subject to criticism. There is no doubt that Heady's work in adapting them to agricultural problems was a considerable advance and that his thinking opened the way to further developments.

Linear programming

One of these further developments deserves special mention: linear programming. This technique arose from mathematical studies by J. von Neumann in the late 1920s on the theory of games and it was further developed by von Neumann and the economist O. Morgenstern. Linear programming was extensively used during the Second World War as one of the essential tools of operational research, that is strategic and tactical decision-making. It is a method for planning the optimal disposition of several resources when the criterion of success is that the combination should result in maximal return or, alternatively, minimal cost. The most useful application of the technique has undoubtedly been in diet formulation for livestock. Given a series of feeds, each containing known amounts of several nutrients and each with a known cost, then linear programming can formulate a mixture of them which supplies the known needs of the animal at minimal cost. The operation can be modified to avoid giving single feeds in excess and the answer of course depends on the prices of the feeds, which can vary from time to time. The computational work is heavy, particularly in real situations when the number of individual feeds to be considered is high, and its use only became commonplace after the advent of electronic computers. Least-cost formulation of diets for different classes of stock is now used generally throughout the feeding-stuffs industry. Many larger farmers, who purchase their feeds in the open market to supplement their home-grown produce, do the same, using commercially available computer programs for the purpose.

Models and their application

The new econometric methods of thinking about farm innovation and the advent of electronic computers led to the development of sophisticated aids to the provision of advice to farmers and also to new approaches to the understanding of the biological processes underlying agricultural problems. Initially, many of the aids were simple and designed to obviate calculation by the farmer. Examples are the ready reckoners for assessing the calving patterns of dairy herds. Later more complex methods of system analysis arose. These had mixed origins. Some arose from developments in general system theory in the United States, others from studies in the mathematics of the spread of infections in populations, still others in general ecological theory; many biologists attempted to build mathematical descriptions to integrate the increasing amount of information that was accruing from the investigation of biological processes. J.E. Vanderplank in the early 1960s pioneered the development of models that predicted the course of epidemics of plant disease and predicted their outcome. Commencing with studies related to crop competition, C.T. de Wit showed how systems analysis could be used in agriculture and the potential value of simulation of crop responses. His work, and that of F.W.T. Penning de Vries and P. Buringh, who worked with him at the Agricultural University of the Netherlands, did much to encourage others to emulate it. He and his colleagues arrived at models that simulated the reactions of crops, throughout their growth, to nutrients, water, weeds, pests and diseases. These and similar models have been used as the basis of advice to farmers on when to take control measures. More dynamic approaches were also made to the problem of estimating the fertiliser requirements of crops by modelling the uptake of nutrients and water from the soil. Here the major work was undertaken by D.J. Greenwood at the National Vegetable Research Station in England. The pioneer in the modelling of the biochemistry of animal growth and milk secretion was undoubtedly R.L. Baldwin of the University of California at Davis. His studies stimulated workers in Australia and in England, notably J. France and J.H.M. Thornley and their colleagues at the former Grassland Research Institute at Hurley, to embark on very large models of ruminal and tissue metabolism. These involved in some instances over 50 simultaneous differential equations, which were solved by numerical integration methods and which attempted to describe the way in which metabolism responded to feed. They were not used in practice in any advisory way but rather served to verify whether hypotheses adumbrated on the basis of specific experiments could be supported in the context of what was known about other aspects of metabolism.

Interest in mathematical modelling of agricultural and biological processes continues to grow. There is now perhaps less concern with formulating massive mathematical models to deal with every eventuality that could affect a crop or to model biological processes at the molecular level when the objective is to provide advice about practical measures that could be taken by farmers. It is perhaps through mathematical resolution in this way that the quantitative information that has accrued from research can best be summarised and put to practical use in the future. The obtaining of the quantitative information in the first place, however, still requires the emphasis on good experimental design and rigorous statistical interpretation which has been characteristic of the developments in the agricultural sciences during the half century of the modern agricultural revolution.

The measurement of substances

While statisticians were grappling with the epistemology of experimental comparison and bringing precision and security to agronomy and animal husbandry, and while computer technology was bringing a new dimension to statistical methodology and the modelling of biological situations in agriculture, the basic methodology of the chemist, whose analytical methods were important tools of the experimenter, scarcely changed. They were still largely based on gravimetric and volumetric estimations, which within their limits yielded results of great accuracy and value but were slow, painstaking and required highly skilled operatives.

But the seeds of revolution were there. In 1903 Russian botanist, M.S. Twsett, separated the pigments of green plants by utilising their different diffusional characteristics and published a number of papers thereafter, which laid the foundations of the modern technique of chromatography (Lederer,

1994). His work was neglected for the next thirty years until methods derived from his concepts began to be used for separating materials available in reasonable quantity. Changes in the analytical procedures which stemmed from this work allowed smaller and smaller quantities of material to be handled and permitted rapid identification and accurate estimation of the quantities of a range of substances which hitherto had proved almost impossible to handle within a reasonable time scale. This released a whole new research capability and made possible the resolution of problems that had previously been difficult to work on because of either the time required or the inadequacy of the technical procedures for the achievement of results. The history of the development of these techniques deserves a book to itself but in shorter form is well described in Lederer's book *Chromatography for Inorganic Chemists* (1994).

Chromatography depends on the repeated subjection of a mixture of chemical compounds to extraction by liquid or by adsorption onto a solid surface. These two different principles of physical separation depend either on differences of solubility in two immiscible phases (partition chromatography) or on differences in adsorption on a solid surface (adsorption or ion exchange chromatography), the method first used by Twsett. This method was the first to be developed at the beginning of our 50-year period for the separation of a variety of pigments and metabolites available in reasonable quantity. The technique is simple and robust: a glass column is packed with one of a variety of adsorbent materials and the dissolved mixture is poured in at the top where the solvent percolates downwards with the materials to be separated moving down the column at different rates and separating, as they go, into bands. The materials can be separated by sectioning the column to remove individual bands or by elution, when solvent is continuously added at the top of the tube and the bands move through the tube and emerge one by one in the solvent at the base.

Partition chromatography, which followed adsorption chromatography, has given rise to many more developments and to a range of very sensitive and rapid techniques; in one form or another it is the basis of most of the analytical chromatography of the present day. It depends upon the fact that dissolved substances are in continual transport between a moving carrier phase and a stationary liquid phase. The distribution of a dissolved substance between these two phases, under near-equilibrium conditions, depends on its relative solubilities in them. The technique uses a porous solid to immobilise the stationary liquid phase; the carrier can be either a liquid or a gas. If the moving phase is a gas a wide variety of substances can be separated as vapours and can be detected by a variety of techniques such as flame ionizing detection, which measures the electrical conductivity of the material in a flame of the carrier gas. It may be used further in conjunction with infrared or ultraviolet spectroscopy and even, with greater expense, mass spectrometry to separate and measure closely related compounds.

Another type of partition system is paper chromatography, where sheets of absorbent paper are used as the means of stabilising the stationary water, or other liquid, phase. The mixture being analysed, often amino acids, carbohydrates, steroids, purines, etc. is placed as a spot on one corner of the paper. Developing solution is then allowed to flow down the paper by capillary action and resolves the initial load (spot) into a number of spots. These spots may be further separated into components by turning the paper through 90° and repeating the process. Paper chromatography has found great use in biochemical investigations of enzyme pathways, etc., where qualitative information can be obtained on the compounds appearing at various points in, for example, a metabolic process, simply and cheaply.

Nevertheless the variant with the greatest potential and value is probably gas chromatography. The possibility of adding to and improving its basic systems has meant that many suppliers of laboratory equipment have come into competition to try to capture the market, and there has been a great development of ever more sensitive, user-friendly, ingenious and robust equipment. It is thus difficult to separate out the contributions of individual workers in the way that was easily clear in the development of statistical methods. But we can point to a number of workers who have made signal contributions in the field. Lederer himself (1994) has had a continuing involvement with the development of chromatography. A.J.P. Martin and R.L.M. Synge published a classic paper (1941) on 'A new form of chromatography employing two liquid phases'. At that time they were both working for the Wool Industries Research Association. In 1947 Martin wrote an article which is one of the best popular accounts of the principles of chromatography and still valuable today. Later, with A.T. James, who for many years led biochemical research at Unilever's Colworth House Research Station, he published a paper (1952) on gas–liquid partition chromatography. Synge went on to work at the Rowett Research Institute, Aberdeen, and at the Food Research Institute, Norwich. Martin worked for the Research Department of Messrs. Boots and for the Medical Research Council. They jointly received the Nobel Prize for Chemistry in 1952.

The availability of these analytical techniques, and the ease with which they could be married to a computing system to provide a Laboratory Information Management System (L.I.M.S.) which was programmed to perform all the tasks of data acquisition including calculation of values and their statistical and graphical treatment, allowed development of extended analytical facilities for agriculture, medicine, etc. Even before the onset of computerisation the technical facility available to research workers in agriculture allowed our understanding of biological processes to advance with great strides in the second half of the fifty-year period of the modern agricultural revolution. Sometimes this was with immediate practical effect, as in reproductive biology and hormone chemistry; sometimes, as in our understanding of photosynthesis and of the resistance mechanisms of plants, the exploitation of the increased understanding is still to come.

In their turn these techniques have stimulated an interest in other new techniques such as the identification of materials within the cell by a combination of the use of radioactive tracers, immunological methods and electron microscopy. They have also made possible some of the developments in nucleic acid chemistry and the whole range of new techniques which were becoming prevalent at the end of our period and which are leading to the present remarkable advances in molecular biology.

The advances outlined here (which represent only a small part of the technical achievement) have revolutionised science so that the experimental methods that gave rise to the modern agricultural revolution have been technically eclipsed. The generation of scientists whose work will fuel the next agricultural revolution already work with a technical armoury that has transformed not only their capacity for achievement but also their scientific philosophy and organisation. With equipment of such power and of such cost, the role of the senior scientist becomes more and more that of a manager and fund-raiser for a number of younger research workers who work with him in a team. The hope must be that in the pressure of technological progress the future Fishers, Headys, Synges, James and Martins of the world will find time to think and room to innovate.

Selected references and further reading

Beaven, E.S. (1922). Trials of new varieties of cereals. *Journal of the Ministry of Agriculture* **29**, 337–47.

Braude, R. (1972). Feeding methods. In *Pig Production, Proceedings of the 18th Easter School in Agricultural Science of the University of Nottingham, 1971*, ed. D.J.A. Cole, pp. 279–91. Butterworths, London.

Crowther, E.M. and Yates, F. (1941). Fertiliser policy in wartime: the fertiliser requirements of arable crops. *Empire Journal of Experimental Agriculture* **9**, 77–97.

Dent, J.B. and Casey, H. (1967). *Linear Programming and Animal Nutrition*. Crosby Lockwood and Son Ltd., London.

Elsley, F.W.H. (1975). *Nutrition of Sows during Pregnancy and Lactation*. Miscellaneous Publications of the Edinburgh School of Agriculture no. 595; Edinburgh.

Finney, D.J. (1952). *Statistical Methods in Biological Assay*. Charles Griffiths and Company, London.

Fisher, R.A. (1925–34). *Statistical Methods for Research Workers*, Oliver and Boyd, Edinburgh.

Fisher, R.A. (1935). *The Design of Experiments*. Oliver and Boyd, Edinburgh.

France, J. and Thornley, J.H.M. (1984). *Mathematical Modelling in Agriculture. A Quantitative Approach to Agriculture and Related Subjects*. Butterworths, London.

Fussell, G.E. (1935). The technique of early field experiments. *Journal of the Royal Agricultural Society of England* **96**, 78–88.

Gower, J. (1985). The development of statistical computing at Rothamsted. *Report of Rothamsted Experiment Station for 1985*, Part II, pp. 221–35.

Greenwood, D.J. (1979). New approaches to forecasting the fertilizer requirements of vegetable crops. *Journal of the Royal Agricultural Society of England* **140**, 101–8.

Heady, E.O. (1952). *Economics of Agricultural Production and Resource Use.* Prentice-Hall, New York.

James, A.T. and Martin, A.J.P. (1952). Gas-liquid partition chromatography: the separation and micro-estimation of volatile fatty acids from formic acid to decanoic acid. *Biochemical Journal* **50**, 679–96.

Johnston, J.F.W. (1849). *Experimental Agriculture.* Edinburgh.

King, J.S. (1927). *Cost Accounting Applied to Agriculture.* Oxford University Press, London.

Lederer, M. (1994). *Chromatography for Inorganic Chemists.* John Wiley, Chichester, U.K.

Lush, J.L. and Hazel, L.N. (1950). Computing inbreeding and relationship coefficients from punched cards. *Journal of Heredity* **44**, 301–6.

Martin, A.J.P. and Synge, R.L.M. (1941). A new form of chromatography employing two liquid phases; 1. A theory of chromatography; 2. Application to the micro-determination of the higher monoamino-acids in proteins. *Biochemical Journal* **35**, 1358–68.

Martin, L.R. (1977). *A Survey of Agricultural Economics Literature.* University of Minnesota Press, Minneapolis.

Neumann, J. von and Morgenstern, O. (1947). *Theory of Games and Economic Behaviour.* Princeton University Press, Princeton.

Nix, J.S. (1984). *Farm Management Pocket Book*, 15th Edition. Wye College, Ashford, Kent, U.K.

Selly, C. and Wallace, D.B. (1966). *Planning for Profit: a Simple Costings System to Help Farmers Meet the Challenge of Falling Prices.* Cambridge University Farm Economics Branch and the British Broadcasting Corporation, Cambridge and London.

'Student' (1908). The probable error of a mean. *Biometrika* **6**, 1–25.

Touchberry, R.W. (1973). The life and contributions of Dr Jay Laurence Lush. *Proceedings of the Animal Breeding Symposium in honor of Dr Jay L. Lush.* American Society of Animal Science, American Dairy Science and Poultry Science Association, July 23rd 1972, Blacksberg, Virginia.

Vanderplank, J.E. (1960). Analysis of epidemics. In *Plant Pathology*, ed. J.G. Horsfall and A.E. Dimond, pp. 229–89. Academic Press, New York and London.

Wit, C.T. de and Penning de Vries, F.W.T. (1985). Predictive models in agricultural production. *Philosophical Transactions of the Royal Society, London* B **310**, 309–15.

Yates, F. and Mather, K. (1963). Ronald Aylmer Fisher 1890-1962. *Biographical Memoirs of Fellows of the Royal Society* **9**, 91–129.

4 Mechanisation

Men, horses and machines

Although the horse was the general source of power on the British farm at the beginning of our period, it must not be forgotten that the steam tractor was an important power source in certain parts of the country 50 or more years earlier. The steam tractor was able to plough the flat fields, sometimes of heavy soil, in the eastern counties using a reversible plough pulled between two tractors at either side of the field. It was also a source of power for the threshing machines, which travelled between farms. However, the main source of power on farms in general was the horse; the most obvious change that has taken place during the course of the modern agricultural revolution has been the disappearance of people actually seen to be working the land. This is of course more obvious in the arable areas. Where formerly there were slow pairs of horses ploughing at the rate of a few acres per week, gangs of workers singling sugar beet day after day, and family parties setting sheaves in stooks following the binder, now there are rapid forays by machines with scarcely a person to be seen. In livestock areas it is much the same. The night and morning routine of the shepherd has been replaced by the quick visit to his flock by the part-time stockman in his Landrover or on his four-wheeled motorbike; gone too are the gangs of milkers to be replaced by the lone, hard-worked dairyman in his clinically efficient milking parlour. The unremitting toil of farm work has been signally reduced by mechanisation until now farm workers deploy more mechanical power than do many workers in conventional industry. Mechanisation of all farm operations, the invention of new machines and devices, and the development of new buildings designed for specific purposes has been central to the revolutionary process.

The decline in the number of people and of the older motive power unit, the horse, during the period from the 1930s to 1980s was commented on in Chapter 2 and specifically in Table 6. These changes are paralleled by the increases in the number of tractors, implements and machines which took place during the same period. There were 532 000 tractors on United Kingdom farms in 1986. At that time there were 234 000 farmers and members of their families working full-time together with 112 000 full-time hired workers: a total of 346 000. In other words there was more than one tractor for every full-time farmer and farm worker. In the 1920s, when the horse was the motive unit of power, the ratio of the number of horses to the number of men and women working on the land was somewhat less than 1.0, suggesting that before the revolution there was less than one horsepower available per person on the land. Obviously, when the horse was replaced, the power deployed per person in the agricultural industry increased considerably. The number of tractors appeared to stabilise in the later 1950s. Their horsepower did not; it continued to increase, and by the 1980s 80 h.p. tractors were common and some tractors had drawbar horsepower ratings of over 200. This was not the only source of motive power on farms. As the revolution proceeded, the larger machines for harvesting grain and roots evolved to become self-propelled. Thus a rough estimate of the power at the disposal of those who worked the land shows that it changed by about two orders of magnitude during the course of the revolution.

It was remarkable how the people concerned with farming accommodated to this change. The blacksmiths soon learned that farriery was a dying art and that the future lay in the repair and maintenance of farm machinery and equipment and in the stocking of spare parts for a wide variety of implements. Those working the land became expert fitters, capable of servicing their complex new tools. Machinery clubs sprang up where information and skills as well as equipment could be exchanged; syndicates for farm machinery arose and here Leslie Aylward, a farmer and agricultural merchant in Hampshire, was a pioneer. The larger farms were soon equipped with well-designed workshops, and smaller farms too had the hand tools, working benches, welding equipment and repair facilities to enable them to undertake the more simple tasks necessary to keep their machines working. The buying and selling of second-hand farm machinery became as interesting a venture as was the trading of horses in the previous era!

Many of the major advances in mechanisation that took place during the revolution were undertaken by commercial engineering firms, both in the United Kingdom and in other countries, particularly in the United States of America, and there were many such firms competing for the expanding markets at home and abroad. It is often difficult to identify which firm made the signal advance and impossible in some instances to single out the individuals concerned. Many of the developments that occurred were in response to other technological changes that were taking place. Particular patents were licensed to other firms in the course of development of a machine. An obvious example of such a response was the development of spraying equipment following the discovery of selective herbicides and pesticides; another was the exploration of methods for drying grain following upon the increased use of the combine harvester. The account which follows deals with developments of mechanisation in a farming sequence – from cultivation to sowing to harvesting of crops – and also with the mechanisation of livestock enterprises and the design of the farm buildings required to deal with technological change.

The need for cultivation

It has always been axiomatic in farming that to grow crops the land must be tilled. Jethro Tull's 'horse hoing husbandry' amplified this by emphasising the importance of inter-row cultivation of crops. This view was overturned by a physicist, Sir Bernard Keen. Keen was appointed in 1913 to undertake studies on the physics of the soil at Rothamsted, and returned there after the First World War and the Gallipoli campaign. The ideas at that time were that moisture was lost from the soil through capillary action. It was thought that surface cultivation broke these capillaries, and thus conserved moisture. Keen overthrew this hypothesis in laboratory experiments and then asked the question 'Is there any value at all in cultivation other than to destroy weeds?'. With E.W. Russell he conducted a series of field trials to ascertain whether tillage was important. He showed that, provided weed competition within the crop was avoided, tillage was not necessary. Excessive inter-row cultivation in the absence of weeds in fact reduced yield, and deep ploughing was quite unnecessary. Keen published a series of papers on cultivation during the late 1930s and early 1940s, culminating in a summary paper, with E.W. Russell, in 1941 in which he gave details of the results obtained over several years. He did not continue this work and after the Second World War became the first Director of the East African Agriculture and Forestry Research Organisation at Muguga.

These revolutionary findings were, perhaps understandably, viewed with a certain amount of suspicion at the time. There was some attempt during the war to reduce the amount of what was called superfluous cultivation such as double ploughing, but no exploitation. Keen's discovery received support from the United States following the calamities of the 1930s. Continued cultivation of prairie land resulted in dust-bowl conditions and called into question the desirability of traditional cultivation. The discovery of the total weedkillers diquat and paraquat in the late 1950s (Chapter 6) changed the situation. These weedkillers would deal with weeds and enable crops to be sown without cultivation (see also Chapter 8, pp. 161–2). Trials soon followed and these were published in the early 1960s, notably by A.E.M. Hood who worked at Jealott's Hill, the agricultural research centre of I.C.I. (now Zeneca plc), to show that grass could be reseeded by killing the old sward and then drilling directly, and that wheat too could be seeded into stubbles in this way. Quick-growing turnips could be planted easily on cereal stubbles to give catch crops, the yields of which were only slightly below those obtained after conventional ploughing. One problem was that the seed drills available did not penetrate the ground very well and wear on the coulters was heavy. New machines were developed very quickly to deal with the problem. By 1972 it was estimated that some 1300 000 hectares were being 'sod seeded' or 'direct drilled' in the U.K. Similar developments were taking place in America, where the technique was called 'no till', and Australia, where it was called 'spray seed'. Despite the high costs of the weedkillers, they were adopted by farmers.

Further development to exploit Keen's finding then occurred with the development of reduced tillage techniques. In these the cultivations for a cereal crop were confined to the surface and ploughing was totally avoided. A central feature was the burning of the stubbles followed by cultivation with spring-tined cultivators to produce a surface tilth. Other implements were adapted to the process, including disc cultivators and rotary cultivators such as the Dynadrive. The critical feature was to achieve a fast work rate. Yields were usually not very different from those obtained with conventional ploughing. This was probably due to the effect of ploughed-in straw in depressing the yield of the subsequent crop: the decomposing straw competed for soil nitrogen. During the 1970s about one third of the total cereal acreage was established without ploughing, entailing the burning of about a million and a half hectares of stubble. The smoke and smuts annoyed local people – including the farmers' wives! The new approaches had an additional merit in that they enabled an easy switch to be made from spring- to winter-sown cereals (especially barley), a change which took place at that time. However, by the early 1980s a move back to ploughing occurred. The reasons for this were largely the compaction of the soil occurring when direct drilling was adopted in wet seasons together with an unacceptable invasion by grass weeds when the practice was continued over several years. What perhaps emerged from the whole of this work was that farmers realised that there were alternative ways in which soil can be prepared for crops, that some cultivations were unnecessary, and that there were ways of taking advantage of small windows in the weather without sacrificing yield. The advent of mechanical spacing of row crops and of selective weed control in them had the effect of reducing the inter-row cultivation so long accepted following Tull's advocacy and so thoroughly discredited by Keen's work.

The tractor and the plough

Despite these excursions into ploughless husbandry, the plough is still the major implement for cultivation. When tractors were introduced before the First World War they appear to have been regarded as mechanical analogues of horses. It was soon learned that, unlike horses, tractors have no innate sagacity. In those early days horse-drawn implements were simply towed by tractors. The difference between the horse and the tractor soon became evident. When a tractor-drawn plough confronts an obstruction the force being exerted does not cease but either results in breakage of the ploughshare or is transmitted, to result in the front wheels of the tractor rearing up with possible injury to the driver. In the same circumstances the horse simply stops! Even with normal ploughing the draught resulted in a slight lifting of the front wheels, making steering more difficult. The solution adopted to solve this problem was to increase the weight of the front of the tractor but this then created problems at the rear, for the wheels slipped. More weight was added to the rear end in an attempt to compensate. There were other problems. When ploughing with horses the ploughman could control lateral movement of the plough and achieve depth control by pressure on the stilts (the handles of the plough); with a plough trailed by a tractor horizontal movement and depth control were slightly more difficult.

The solution to these problems was found by Harry Ferguson and his designer, William Sands. Ferguson, the son of an Ulster farmer, was apprenticed to engineering in his brother's car repair workshop in Belfast. He designed, built and flew the first aeroplane ever to be flown in Ireland and, during the First World War, was engaged in the promotion of tractors as part of the food production campaign. This sparked his interest in the mechanisation of agriculture. The outcome, patented over the period from 1925 to 1938, was the 'Ferguson System'. In this implements were made integral with the tractor; the linkage of the plough to the tractor was arranged to ensure optimal downward forces on all four wheels of the tractor and minimal lateral movement, while depth control and hoisting of the implement at the end of a bout was achieved by hydraulics. In addition a new tractor was designed; production of the Brown–Ferguson tractor, the prototype of the 'little grey Ferguson', began in 1936. After some famous litigation, an agreement was then reached with Henry Ford in the United States to result in the Ferguson system being applied to the Fordson tractor. It was the invention of this system that led to accelerated adoption of the tractor as a motive unit for all farm operations.

The tractor, even at this time, was regarded as a mechanical horse to work in the fields. But the horse did more than this; it also undertook most of the farm transport. The steel wheels of the tractor and the protruding strakes on them caused damage to roads. This was partly overcome by fitting steel bands over the lugs; this, incidentally, caused sufficient slip to prevent traction in some instances. The ultimate solution was to use pneumatic tyres; these were introduced by Firestone in the United States and taken up by other tyre manufacturers. The take-up was, however, very slow during the war years. Italian firms did considerable work on the shape of the treads on these tyres and optimal pressures, and later introduced radials.

Ploughs, rotary and powered implements

There was relatively little advance in plough design during the period except for an increase in the number of furrows to be commensurate with the higher draught of the tractors being employed and an increased emphasis on one-way, reversible ploughs. The first rotary cultivator was designed in 1912 by an Australian engineer. A.C. Howard from New South Wales. He came to England in 1938 to form a company, Rotary Hoes Ltd, with E.N. Griffith, who was farming in Essex, and in which firm A.C. Howard's son, J. Howard, continued design work. The Howard rotavator has been much used for the production of seed beds and was extensively used to break up permanent pasture during the Second World War. Powered harrows were developed in the 1950s to accelerate the production of seed beds with a large number of patents granted, particularly to Figlo Ltd.

Seed drills

Several firms developed new types of delivery mechanisms for seed drills to replace the time-honoured cup arrangement; several of them used hydraulic delivery. What was perhaps most interesting was that the width of the drills for most crops had changed very little before the revolution; they were dominated by the size of horses' feet. Work undertaken in horticulture, notably by J.K.A. Bleasdale at the National Vegetable Research Centre, Wellesbourne, questioned existing spacing of commercial vegetables, to advocate reduction. A number of new seed drills were produced for this purpose and a new exploration of possibilities for arable crops was made. Stanhay Ltd. and A.C. Bamlett Ltd. were in the forefront of these developments and in that of combined fertiliser and seed delivery, which also became commonplace.

The seeding of sugar beet is of interest. The sugar beet 'seed' is a cluster of fruits which in the corky bark holds two or three seeds. Before the Second World War the germinated seedlings were separated by hand hoeing and later a technique of cross-blocking was employed in which the crop was machine hoed at right angles to give small tufts of germinated seed, which were easier to hoe by hand. To economise on labour, attempts were then made to reduce the number of seed in each cluster by rubbing it and grading it. Damage to the seed occurred. Gravity separation of remaining fragments then took place, which also removed non-viable bits of the true seed and produced some economy in labour. In 1950 beet plants were discovered in the United States which had only one true seed per cluster; by 1965 they had been introduced into the United Kingdom and the propensity to produce a single germ had been incorporated into British varieties. Single-seeders were developed by the agricultural machinery industry to deliver what was a highly standardised product and by 1974 about a quarter of the British sugar beet crop was grown without the use of hand labour (Chapter 7).

Mechanisation of harvesting

The first combine harvester for grain in the U.K. was a McCormick Deering 10 ft machine imported from America in 1927; it was joined the following year by a Massey Harris with a shorter cutter bar. Progress – all by importation of North American machines – was slow. Only 33 machines were in use by 1933 and they were not satisfactory. Designed for the drier conditions of Canada and the United States, they could not deal with the tall straw of the time or with the heavy stubbles. The British crops were also dirty with weed seeds, which had to be removed, and the grain emerging had a high moisture content. The latter problem was overriding; many attempts were made to design farm grain driers. The first combine to be built in Britain was that designed by D. Bomford; all that it did was to demonstrate that American design was better than British! A smaller combine with a 6 ft cut was introduced in 1935 and this did more than anything to create interest. Subsequently American engineers began to take the market seriously and, after the Second World War, so too did continental firms. There remained many problems related to drying of grain, for maltsters avowed that grain could not be dried on farms without affecting its malting quality. Considerable investigational work was done by the National Institute for Agricultural Engineering to find the principles involved in drying grain; the Institute developed electronic meters to monitor continuous flow driers as well as low-rate airflow meters, marketed in 1952, for use with ventilated bins. T. Ensor, a farmer, did very useful work in developing floor methods of drying grain, for which he received the Massey Ferguson Award in 1967.

As with the combine harvesters, the baler was an American introduction, although plans had been made by A.J. Hosier to build one here. The American machine imported had a low rate of work and the tying of the hay bale was with wire; this was changed to heavy-duty twine and, recognising the importance of maintaining a low moisture content in the material to be baled, very good hay was made using both stationary and pick-up balers. The conveying of hay and straw to the stationary baler was by tractor sweep and the transport of the bales to store was by buckrake, the implement invented by Rex Patterson, a distinguished farmer. The 'Big Baler' was invented by two British farmers in 1972 – P. Murray and D. Craig – and was quickly accepted by farmers. Farmers had many alternatives to choose from; the number of makes of baler being marketed in 1986 was over 20.

Before hay is made it has to be cut! In the 1930s this was entirely by the finger-bar mower with its triangular reciprocating knives. These changed in the 1950s to rotary mowers and later to disc mowers since they had lower power requirements.

Most of the machinery for sugar beet harvesting was European in origin but a notable British machine was built by A. Standen and Son Ltd. For potatoes there was, however, a considerable and highly sophisticated investment to be made in machinery towards the end of our fifty-year period, as electronic discrimination became possible between tubers, clods and stones. Machines were also invented for removing stones from the ridges of soil before potato planting.

The plethora of machines

Like most people, farmers do not like paying taxes; during the mid- to late 1970s their incomes were reasonable, particularly for those in certain parts of the country who had grown potatoes in the poor harvest year of 1975. In the following winter and spring, farm gate prices for potatoes rose threefold. Investment in machinery was very heavy at that time – not for necessity but to take advantage of tax concessions and to dispose of cash. And there was considerable choice available. The farming periodicals contained many advertisements and, in addition, several pages of articles describing new implements and machines. The great international agricultural shows were almost entirely dominated by demonstration stands of farm machinery together with considerable promotional literature. Even the smaller local agricultural shows gave prominence to machines and implements. With this pressure it is perhaps understandable that many farms became overstocked with farm machinery, and as this situation dawned on some farmers it gave an extra impetus to them to purchase more land so as to spread machinery costs. With higher general costs in the early 1980s, some farmers indeed dispensed with all but the minimum mechanical power for transport within the farm and contracted out arable operations to a growing number of independent contractors. Certainly, towards the end of the revolution, most major drainage work and lime spreading was done by contract using very heavy machines specifically designed for the purpose.

Manure disposal

A new departure in farm mechanisation was in the disposal of effluent from intensive animal enterprises. In the 1930s the disposal of manure was on a relatively small scale because the number of stock kept on a holding was small and the work was done by hand. The advent of the front-end loader helped to take the hard manual effort from the task but, as the size of enterprises increased, so schemes that verged on civil rather than agricultural engineering came into being to cater for farms with 300 or more cows and a thousand pigs. Large lagoons were built to accommodate effluent, with aeration equipment to aid its oxidation. Pumps and slurry tankers became commonplace to dispose of what was regarded by conservationists as a hazard to the countryside. A further development at this time was mechanical equipment; first to clean livestock buildings of excreta and secondly to separate the semi-liquid from the semi-solid components to ease the disposal problem. An interesting development was in the anaerobic fermentation of slurry to produce methane gas. This was an old process much used in the Far East, and also by some municipal authorities in the United Kingdom. Agricultural versions were devised, notably by P.N. Hobson of the Rowett Research Institute, to link with internal combustion engines which provided ventilation and other power to livestock buildings in summer and heat in winter. Hobson set up a consortium of manufacturers to promote their adoption but with a fall in the price of other fuels the take up was smaller than had been anticipated.

Farm buildings

Following the dereliction of the late 1920s and early 1930s, the reduction of agricultural buildings during the Second World War and the shortage of materials thereafter, little building was undertaken to provide the basis for advance in farm practices until the 1950s. Part of the Government support was channelled to help the provision of adequate buildings on the farm, and at that point there was a considerable expansion in the pace of farm building. Much of the advance came from the suppliers of buildings; considerable work was carried out by the National Institute of Agricultural Engineering at Silsoe, by the Scottish College's Farm Buildings Investigation Unit at Aberdeen, and by the advisory groups countrywide. An important development at this time was the advent of the single-truss roof. The importance of lightweight trusses, which could hold up a wide roof without additional support, lay in the unimpeded floor space they offered. Where the building was also high enough, and with adequate doors, tractor work became possible throughout the whole area and trains of, for example, potato-sorting machinery could be positioned. Grain could be stored and a variety of other uses were facilitated. The flexibility this construction gave was invaluable. Similar buildings could be devoted to animal production with simple penning and feeding arrangements easily incorporated in the design and just as easily removed if necessary.

For pigs and poultry more specialised buildings were required; here, emphasis was placed on fairly lightweight external construction, high insulation and always the linked process of convenient effluent disposal. Cognisance had also to be taken of the physiological requirements and temperature regulation of the pig and fowl. For pigs, outstanding work was undertaken by L. Mount and his associates at the Institute of Animal Physiology. They determined the heat loss by pigs of all ages by radiation, convection, evaporation of moisture, and conduction to the floor, so as to provide designers with information for planning pen design. Similar studies based on single growing birds were of course made, notably by A.R. Sykes at Wye College, but designs of poultry houses were mostly arrived at by building them to different specifications and monitoring the output of eggs or broiler carcasses.

Housing and milking the dairy cow

Most of the milking of cows in the 1930s was by hand. Despite the invention of the milking machine, herd size in general was not sufficient to warrant the cost of the plant. The cows were tied up in winter and fed and milked at their permanent standing place within the byre. During the years that followed a variety of ways were devised to reduce labour in cow care; these used open yards or cubicles where the cows could roam at will and from whence they moved to a specially designed parlour for milking by machine. These parlours were of diverse construction. In the earlier models the cows stood side by side in a range of stalls, then the stalls were raked to form a herring-bone system to facilitate access of the dairyman to the udder. Later the floor of the dairyman's working area was lowered relative to the cows by constructing a pit about 30 in (75 cm) deep behind the cows. More advanced designs saw sets of herring-bone stalls arranged in parallel, back to back, and in triangular fashion (the 'Trigon') or arranged as the outline of a polygon (the 'Polygon') (Culpin, 1992). After that, rotary parlours appeared, in which the cow stepped into a stall on a rotating platform, which revolved at a rate sufficient to ensure that the cow was milked before arriving at the exit. However, by 1986 (and still today) the majority of cows were milked in herring-bone parlours with the cows raised above the dairyman for easy udder washing, teat treatment against mastitis and attachment of the milking cluster. These systems allow one person to milk a surprisingly large number of cows, as many as three hundred in the 'Polygon', but at these levels problems of boredom and mental fatigue become as important as the physical ease with which the operation can be conducted.

The firms of Gascoigne in Britain and Alfa-Laval in Sweden were mostly concerned with these developments and just as importantly, with the development of equipment for automatically weighing milk output, recording the input of concentrated feed, automatic cluster removal and efficient circulation of cleaning solution through the pipes. Development work on the cleaning of pipe systems at the National Institute of Agricultural Engineering at Silsoe led to the Acidified Boiling Water Method; Gascoigne developed the use of injected pressurised air to aid the cleansing process in large-bore pipes. The milk so produced was stored in refrigerated bulk tanks (sometimes following passage through a heat exchanger). Further sophistication came with the use of transponders hung round the cow's neck, which allowed the collection of information on daily yield and concentrate intake in individual cows; with the use of computers, the dairyman was enabled to introduce additional biological information about individual cows and to keep a close management oversight of the enterprise. Needless to say the range of options for the installation of new milking systems now available means that choice to combine as many useful attributes as possible becomes more and more sophisticated.

Just as with the development of the tractor and modern tractor-powered implements, it is important to note here that, although there was a valuable input by the scientists at the National Institute of Agricultural Engineering and the National Institute in Research in Dairying (Thiel and Dodd's work is described in Chapter 11), much of the development push came from the industry and its service industries. The Agricultural Advisory and Development Services monitored, rationalised and objectively tested the systems to help the development forward in a smooth and sensible fashion.

Selected references and further reading

Allen, H.P. (1981). *Direct Drilling and Reduced Cultivation*. Farming Press, Ipswich.

Cashmore, W.H. (1975). *A Career in Agricultural Engineering*. The Massey Ferguson papers.

Cox, S.W.R. (1984). *60 Years of Agricultural Engineering*. National Institute of Agricultural Engineering, Silsoe, Bedfordshire.

Culpin, C. (1992). *Farm Machinery*, 12th Edition. Blackwell Scientific Publications, Oxford.

Fraser, C. (1972). *Harry Ferguson, Inventor and Pioneer*. John Murray, London.

Gibb, J.A.C. (1988). *Agricultural Engineering in Perspective*. Institution of Agricultural Engineers, Silsoe, Bedfordshire.

Keen, B.A. and Russell, E.W. (1941). Studies in soil cultivation. *Journal of Agricultural Science, Cambridge* **31**, 326–47.

Pereira, H.C. (1982). Bernard Augustus Keen. *Biographical Memoirs of Fellows of the Royal Society* **28**, 203–33.

Quick, G.R. and Buchele, W.F. (1978). *The Grain Harvesters*. American Society of Agricultural Engineers, St Joseph, Michigan, U.S.A.

Williams, M. (1980). *British Tractors for World Farming, an Illustrated History*. Blandford Press, Poole, Dorset.

5 *Soils, fertilisers and water*

Background

This chapter illustrates another way in which science influences agricultural practice. Here there are few immediate, transparent connections between discoveries and new profitable practices. Instead there is a gradual and painstaking accretion of knowledge to provide a framework for the thinking of practical men and their advisers, to allow other work to be understood (cf. Chapters 6, 7, 8 and 9) and decisions to be taken in a rational and safe context.

Soil is the very foundation of agriculture. On it grow the plants needed as food for human society and its animals. The majority of soils derive from rocks and drifts, under the influence of climate and vegetation; they encapsulate, for the knowledgeable observer, centuries and even millennia of their history. Individual soils develop over long periods of time. Bare rock surfaces are comminuted by the action of water entering cracks and as a result of temperature changes the interspersed water breaks up the rock into smaller and smaller fragments. Drift material carried by wind, water or ice and deposited at a distance from the parent rock has already undergone the early stages of soil formation. The comminuted rock is further sorted, under the influence of surface water, into grades, with the smallest fragments collecting at the bottom of slopes. The process can be seen in miniature on any roadside cutting through soft rock. On a grander scale the same process leads to catenary sequences of soils down a valley slope to produce a series of soils. In these cases not only do the particles move, but also, in time, the chemical elements released by weathering are carried to the deposit, further emphasising the differences between soils in the chain. Soils that are transported by rivers or glaciers may contain rock fragments from widely separated geographical areas and be derived from a mixed source of parent materials; in contrast, sedentary soils are derived mostly from a single rock material. Comminuted rock is not mature soil; at an early stage it is colonised by adapted plants (algae, lichens, mosses and liverworts) to be followed rapidly by pioneering ferns and flowering plants, when the process of soil formation proper begins, as humus, a breakdown product of the lignin of vascular plants, is added. The type of soil that develops in a particular situation is determined by the nature of the rock or drift material, the topography, the vegetation that develops and the climate. The climate has a preponderant effect in differentiating soil types; dissimilar rocks under a single climatic regime may give rise to very similar soils.

The recognition of world groups of climatic soil types as the primary categories of soil classification was one of the main advances in pedology up to the beginning of our revolutionary period. The reason why climate is of paramount importance is that two elements of climate – water and temperature – are, as outlined above, important factors in soil genesis. The climatic soil types are not uniform but show variations according to their parent material, topography, drainage and the available flora; all of these factors interact.

Chemical weathering of the rock fragments, which gives rise to the mineral portion of the soil, takes place more rapidly at higher temperatures, so that we have tropical soils often weathered to great depths. The balance between percolation and evaporation of the incoming rain is markedly affected by climate and determines the rate of leaching of the chemical elements of the soil. When the soil is acid or neutral in reaction, carbon dioxide dissolves in the soil water (to form carbonic acid); when percolation is high the calcium, iron and other cations, as well as parts of the humus fraction, are washed through the soil and sometimes deposited at lower levels in the soil as, for example, iron or humus–iron pan (podzol). The nature of the vegetation colonising the soil is determined, among other things, by the base status of the soil, by its permeability and by the ambient temperature. The contribution of the vegetation to the soil, in death, is similarly determined. In acid, wet, cold conditions, an acid humus layer is built up which contributes to the soil mainly through its part in the leaching and podzolisation process. In warmer, drier, less acid situations the plant remains decay to a form of humus, a dark colloidal material derived from the breakdown of lignin, which is mixed intimately with the mineral matter from the chemical weathering of the rock to form the reactive material of the soil.

The end results of these processes can be classified as a number of soil groups (such as brown earths, podzols) made up of soil series often named after the district in which they were first recognised and containing defined soils varying in texture, chemical composition, colour and depth. They show in sections, achieved by digging soil pits, layers of recognisably different materials. The sectional appearance is referred to as the soil profile. A very clear account of these soil-forming processes is given by Simpson (1983).

Taxonomy

Enlightened by some later thinking, this outline of our knowledge of soils demonstrates the extent of our understanding in 1936 when, conveniently for us, G.W. Robinson published the second edition of his text-book entitled

Soils, their Origins, Constitution and Classification. What followed was first of all an attempt at systematisation and clarification of knowledge of soil taxonomy. It had begun, a little earlier, with a group of Russian workers led by V.V. Dokuchaiev, and resulted in the publication of *The Great Soil Groups of the World* by Konstatin Dimitrievich in 1927. In 1950, W.L. Kubiena published *The Soils of Europe: Illustrated Diagnoses and Systematics* on similar principles. A different methodology had been established initially in the U.S.A., where a Soil Survey Department had been established in 1899. Like much of the American agricultural research of the time it was directed towards providing immediate help to the farmer. The methodology was based on local needs. The soils were classified into Soil Series, named after the districts in which they were initially found, and further subdivided in terms of texture into Soil Types. Attempts were then made to integrate the Russo-European methodology with the American, particularly by C.F. Marbut, who presented his system to the First International Congress of Soil Science in 1927. This did not achieve immediate recognition but encouraged many workers to contribute ideas. Some quite serious etymological flaws in the proposed nomenclature appeared during this time of evolution but they were gradually ironed out until by the end of our period a system had been constructed which received universal acclaim and achieved world recognition. Its adaptation to British needs was achieved by much painstaking work over the period and was finally brought together by Avery (1980).

Throughout the period the United Nations Food and Agriculture Organisation took on more and more of a coordinating role in these matters; a *Map of European Soils* was published under its aegis in 1960. More maps for different parts of the world were to follow.

Activity in soil science increased with developments in universities and colleges of agriculture and in government agencies devoted to the subject. In the U.S.A. the Soil Conservation Service of the U.S.A. had an increasing responsibility for the development of surveys, not only of soil classification but also for land use capability, types of agriculture and climatological interfaces with agriculture and horticulture. In the United Kingdom, national Soil Survey Departments were established in 1939, building on the earlier efforts of individual advisory soil chemists. They became the responsibility of the Agricultural Research Council in 1946. Two of the founding figures in soil surveying in the United Kingdom were William Gammie Ogg, in Scotland, and Gilbert Wooding Robinson, in England and Wales. Robinson became Advisory Soil Chemist at Bangor in 1912. On his own initiative he attempted a soil survey of the surrounding area and found that he could establish no rational system if he relied on the geology for his primary classification. Nearly identical soils could be derived from Cambrian, Ordovician or Silurian rocks. Soil texture proved equally frustrating as a classificatory criterion. W.G. Ogg spent time with the American Soil Survey in 1919 and became convinced of the value to be derived from soil surveying. In 1924 he began a survey of East Lothian, Scotland, in conjunction with J.O. Veitch of Michigan and J.B. Simpson, using the profile method, for the first time in Britain. In the next few years the profile method became the accepted method of surveying and under the leadership of Robinson and Ogg was adopted on a uniform basis and integrated with efforts in other parts of the world. Ogg became Director of the newly formed Macaulay Research Institute at Aberdeen in 1930, and at an early stage Alexander Muir became Head of the Scottish Soil Survey, established there. As noted above the English Soil Survey began under Robinson at Bangor in 1939 and later was transferred under the leadership of Alexander Muir to Rothamsted Experiment Station, where Ogg was now the Director. The interest in soil science in Britain was quickly transferred to the British colonial territories, where a number of distinguished soil scientists (Hardy, Greenwood, Nye, Beckett, Tinker, Charter) made contributions, some of them returning to the U.K. to continue their work. Only some of them were engaged in soil surveying in a strict sense. The organisational difficulty of some of this work in other countries was considerable. Cecil Charter, when surveying in Ghana, used to organise the cutting of pathways through the vegetation (including monsoon forest) at distances of two kilometres with sampling stations at intervals of two hundred metres across many kilometres of difficult terrain, with more intensive surveying in areas of variable soil. The whole operation looked more like a military manoeuvre than a scientific expedition.

Chemistry and plant nutrition

The provision of a secure scientific framework of knowledge of the taxonomy of soils cleared the way for two further developments. These were in no way related to the taxonomy of the soil, but identification of soils helped communication between scientists and some of the understanding of the soil-forming process, derived from attempts at identification, helped the formulation of experimental work. The first was the study of the chemical and physical processes which took part in the soil's continuing development, the end effect of which was to provide nutrients for the plants growing in it. The second was the study of its microbiology, of equal importance.

Much had been learned empirically about the way the soil contributed to the nutrition of plants, and the work of the nineteenth-century chemists, epitomised by Liebig, had given scientific plausibility to practical understanding. Others such as Dyer (1894) opened a way to an understanding of the availability of minerals in the soil, work which was later developed by E.G. Williams and J. Tinsley. Gilbert and Lawes, from an entirely different standpoint, had shown how plants respond to manures applied to the soil, in the experiments summarised by Daniel Hall (1905) but comparatively little was known about how the chemical elements important for the plant behaved in the soil.

It is perhaps surprising that the rapid development of analytical chemistry in the early part of the twentieth century did not result immediately in the establishment of an adequate methodology for the chemical evaluation of soils. In fact it had to wait upon an increased understanding of the source of plant nutrients. Three sources were demonstrated from which the nutrient ions become available at the root surface: the soil solution, the exchangeable ions, and the readily decomposable minerals.

The soil solution can be isolated by displacement from the interstices of the soil by a variety of methods. To make sense of the results, the condition of the soil at the time of collection has to be carefully specified because the substances in solution vary with the acidity and the moisture content of the soil. Of the ions important in plant nutrition, nitrate concentration varies with the soil moisture content whereas the concentrations of potassium and phosphate ions do not, for quite complex chemical reasons. The movement of plant nutrients in the soil has been the subject of study by many plant physiologists and soil scientists (Nye and Tinker, 1977) providing a quantitative view of the movement of water and dissolved ions to the root surface and into the plant root. In the soil this may be either by mass flow as the solution around the absorbing root is depleted by translocation, or by diffusion down a gradient to the root surface. Controversy continues between the protagonists of these two systems but in practice we can think of both processes occurring together.

The depletion of ions from the soil solution by plants in full growth is rapid. The replacement of nutrients is now understood to be in part through the ion exchange mechanism. Humus, produced by degradation of the lignin in decaying plants, and clay minerals, produced by weathering of the rock debris, carry negative charges on their surfaces which are balanced by cations, typically calcium, magnesium, potassium, sodium and sometimes aluminium and manganese. These are held on the minerals and are in equilibrium with their counterparts in the soil solution. Depletion of nutrient ions from the nutrient solution, the dilution or concentration of that solution, variation in the pH of the solution, or addition of material from the decay of previous vegetation or farmyard manure, can all affect the balance of cations available to the plant. Many individuals have contributed to our understanding of this phenomenon but among British workers, Schofield, Beckett and Tinker should be mentioned. Not only does an appreciation of the cation exchange mechanism help the rationale of fertiliser application but the discovery of the part played by aluminium as an antagonist of growth in certain plants, and of its release by the cation exchange mechanism in acid conditions for absorption by the plant, provides a rationale for lime applications for crop plants. Complex compounds containing aluminium are constituents of some of the clay minerals and detached aluminium cations are present in soils derived from them. In alkaline conditions the aluminium is precipitated on the surface of the clay as gelatinous, simple and more complex hydroxides and the cation exchange mechanism is suppressed. In acid conditions the aluminium ions are released into the soil solution and become available to plants.

The availability of aluminium in acid conditions is also thought to influence natural vegetation. Calcium- and magnesium-rich soils are well known to carry floras rich in species. In contrast acid soils carry an impoverished flora of a limited number of apparently specialised species. It is now considered possible that the availability of aluminium in the acid soils, where it accumulates in some plants to a concentration that causes poisoning, may be in part the determining factor in the vegetation pattern. As in all situations where there are a number of factors that inter-relate, caution is required in interpretation.

Well before 1936 it had become clear that certain metals were required in very small quantities by plants for normal growth. Thus Kathleen Warington (1923) at Rothamsted had shown that boron was needed for the growth of broad bean (*Faba vulgaris*) and other plants and American work stretching forward to 1939 showed that copper, molybdenum and zinc were important for potatoes and fruit trees. Indeed the work on copper by Allison, Bryan and Hunter (1927) is a classical study, which released large areas of previously agriculturally worthless land derived from the raw peats of the Florida Everglades for cultivation. Thus a class of elements not normally supplied with the usual fertiliser treatments came to be recognised as the 'trace elements'. They are probably always important components of enzymes. The availability of a number of these trace elements is also influenced by the acidity or alkalinity of the soil. Boron, which accumulates on the soil organic matter, becomes unavailable in alkaline soils particularly in dry seasons where microbial mineralisation is prohibited. Manganese is abundant in many soils and weathered as the manganous ion which is available to the plant. In slightly alkaline soils it may be converted by bacterial action to the manganic form, which is not available. In very acid soils it may become so abundant as to cause toxicity problems for plants. Iron, which is abundant in most soils, may also be converted to the insoluble ferric form by bacterial action in well-drained, base-rich soils. Molybdenum is taken up by the plant as the molybdate ion, which in acid soils is rendered unavailable by reaction with iron compounds. When molybdenum toxicity arises the reason is the high molybdenum status of the base material of the soil. Cobalt is closely associated with manganese and becomes unavailable when the manganese is oxidised to the manganic form in base-rich, well-drained soils. The complexity of the situation is well demonstrated by manganese deficiency, which is an important problem in cereals on some of the lighter soils as well as being important in potatoes on the organic soils of, for example, the English Fenland. Passioura and Leeper (1963) showed that compacting a light soil reduced the manganese deficiency of oats, and Holmes (1980) and his co-workers, impressed by the remarkable strips of healthy plants observable in barley crops in the areas of compaction under drill and tractor wheels, found an interaction with fertiliser placement and type such that manganese deficiency was reduced in the presence of acid fertilisers placed close to the roots or brought closer to the growing roots by rolling or other compaction.

Not only is there an active cation-exchange mechanism but soils also contain free positive charges on inorganic substances rich in iron and on basic groups on the soil organic matter. To balance these, anions are captured and enter into an exchange situation; the detailed mechanism is still little understood. Important anions for the plant are orthophosphates, which can move in and out of the soil solution from a loose combination with iron silicate minerals. As well as this situation where the phosphate anions are mobile (which parallels the exchange of the metallic cations), phosphates may be permanently held and are then said to be 'fixed' by soils so that a large proportion of soil phosphate is not at a given time available to plants. Simpson (1983) claims that more than 90% is unavailable; some of the remainder is available 'with difficulty' and only about 1% is available to the plant. Care has therefore to be taken in the use of phosphate fertilisers: if applied in excess of the plants' needs, they will be 'fixed' by the soil and wasted. Phosphate introduced to soil as monocalcium phosphate is soluble in water and initially available to the plant. In acid soils phosphates become unavailable owing to the presence of the large quantities of aluminium and iron hydroxides (insoluble) with which the phosphate combines. In the pH range 5.5–6.5, which is a good range for other husbandry purposes, fixation of phosphate is at a minimum. In calcium-rich soils above pH 7 other insoluble phosphates of calcium are formed, making the phosphate unavailable to the plant.

Sulphur, required by plants as an important part of many amino acids, has not been closely studied in the soil. In British soils it has not been in short supply until recently. It is taken up by the plant as sulphate, which is available from the mineralisation of organic material and from the breakdown of sulphur-containing rocks. In addition at the beginning of our fifty-year period it was constantly supplied to the soil by the pollution from coal-burning factories and power stations. In addition a good deal of the nitrogen supplied in fertilisers was ammonium sulphate (originally a by-product of coal gas production). More recently the attempts to reduce pollution and change to ammonium nitrate and ammonium phosphate for use as fertilisers has led to certain areas of the country beginning to show deficiency symptoms, so it is likely that before long the provision of sulphur will be required as a routine fertiliser treatment. Selenium is found in close association with sulphur in certain rocks and can be absorbed by plants, substituting for sulphur in amino acids. Some crops are selenium accumulators (e.g. lucerne) and if the quantity becomes too high, stock feeding on the plants will be poisoned. Equally, selenium may be in short supply so that animals feeding on herbage become deficient in selenium and suffer from white muscle disease (Chapter 9).

Nothing has been said here about nitrogen, which as nitrate is the dominant fertiliser requirement of the plant. The reason is that the material has a relatively short life in the soil, being taken up rapidly by plants (Tinker, 1979) or by microoganisms. Its return from dead organisms – mineralisation – is the basis of many husbandry systems and is also the frequent cause of release of nitrate into water systems and into drinking water. Little work was carried out on the behaviour of nitrogen in the soil in the first half of our period; later Gasser (1964) worked hard to solve the problem. Since 1975 there has been an explosion of work on the problem, aided by the advanced technology which has become available (Smith, 1983).

The important organic fraction of the soil has also proved singularly recalcitrant to experimental investigation. It has been clear to practical plantsmen and soil scientists for generations that the development and maintenance of humified organic matter in the soil confers a range of benefits. As Simpson (1983) says 'it will help greatly in the prevention of cultivation pans, surface compaction, frost heaving, clod formation, wind erosion, capping and puddling, drought, low nutrient status and the excessive leaching of lime and fertilisers. It will also assist in the removal of excess water from the surface soil, the warming up of soils in the spring and will help to create a type of soil structure in which the fine feeding roots can ramify in search of nutrients. This is a formidable list of benefits'. During the period 1936–1986, methods for handling the material experimentally have evolved and work by Jenkinson, Bremner, Swift, Parsons and Cheshire have begun to clarify its formation and role in the soil (see also E.W. Russell, 1973).

Microbiology of the soil

Microbial activity in the soil plays an important part not only in the reduction of plant and animal debris into less structured material, which is eventually incorporated, but also in the cycling of nitrogen and minerals. At the beginning of the century Winogradsky, Beijerinck and Wharton had identified the bacteria that fixed nitrogen in the soil. Later work demonstrated a denitrification process in soils low in oxygen; however, the general role of microbes in the soil had not been studied in detail. Much of the work in the period with which we are dealing has been devoted to the elucidation of the ecological relationships of the organisms in the soil. A picture has emerged of a three-dimensional, below-ground ecological distribution disturbed only by the passage of roots, which create a series of ecological niches on their surfaces, and by the localised successional ecology on recently incorporated organic substrates (Burges, 1958).

Two important concepts derive from this work. First, the microorganisms are in an equilibrium such that an exotic species introduced into the soil finds no nutritional niche and is soon eliminated (Parks, 1955, 1957). Second, some at least of the interaction between organisms derives from their ability to produce antibiotic substances (Waksman, 1947). The history of penicillin, its discovery by Fleming and its development by Florey and Chain, is well documented. Although *Penicillium chrysogenum*, from which it was derived, was not isolated from soil, it is in fact a member of the soil mycoflora. Penicillin is an antibiotic which, with its derivatives, has contributed greatly to the control of animal and human disease. It is, however, active only against Gram-positive bacteria. It was the isolation by Waksman in 1944 of streptomycin from the soil-inhabiting actinomycete *Streptomyces griseus* that demonstrated that antibiotics active against Gram-negative bacteria were available. Here the material was particularly effective against the tuberculosis organism. This discovery set off a frenzy of work on the isolation and testing of the antibiotic powers of soil-inhabiting organisms, and produced a great many new antibiotics, several of which later became important implements for disease control in humans and animals (Chapter 11). In 1947 chloramphenicol was isolated from *Streptomyces venezualae*. It was a genuine broad-spectrum antibiotic, active against *Rickettsia* as well as *Salmonella* and other organisms, hitherto difficult to treat. The chemical evaluation of the antibiotic substances found in nature, and their synthesis and chemical modification, has produced more and more powerful materials for therapeutic use.

The simple concepts of a microbial soil flora with its own ecology allowed the development of other important ideas. Garrett (1956, 1981) working with *Gaeumannomyces graminis*, a soil-borne root disease of cereals and of importance world-wide, found that the fungus could not live in the soil and depended on a food base, which it had previously colonised, to enable it to survive over winter and infect the roots of winter and spring cereals. Such behaviour suggested that the fungus was not equipped to compete directly with the normal soil microbial flora. From this beginning Garrett developed a theoretical treatment of soil-inhabiting pathogenic fungi which has helped generations of plant pathologists to understand the fungi with which they were working.

Another aspect of the soil microbial flora interacting with plants is found in the symbiotic relationships between vascular plants and beneficial root-inhabiting organisms. There are two categories: the first is the mycorrhizal relationship, when fungi inhabit roots and take part in the nutrition of the green plant from the soil; the second is the bacterial relationship with the roots of legumes to form root nodules.

Mycorrhiza

The mycorrhizal habit, a symbiotic relationship between plant roots and fungal partners, is very widespread. The vesicular–arbuscular form can be found in the roots of a wide variety of plants world-wide, but not apparently in all plants all the time (Mosse, 1963). The sheathing mycorrhizas of the roots of forest trees are more or less co-extensive with their hosts. Two other mycorrhizal associations are nearly always present in the Ericaceae and the Orchidaceae. Their interest for agriculture and forestry follows from early claims that they increased the growth of plants and were absent from some soils on which their hosts consequently failed to grow. There is no doubt that, in soils that are below the level of fertility normally associated with agricultural production, mycorrhizal infection can increase the uptake of nutrients from the soil. The mechanism has been explained by Mosse (1963) and Harley (1959) and the evidence is thoroughly discussed in Harley and Smith (1983). The situation can be simply pictured as the provision of carbohydrates by the plant host, which allows colonisation by the fungal symbiont. An enhancement of the root surface area by the extending fungal hyphae allows the exploration of a large volume of soil from which the nutrient in short supply is accumulated. This is often phosphorus. There is no evidence that under normal levels of agricultural fertility the mycorrhizal infection contributes significantly to the nutrition of the host. However, the possibility of benefit under agricultural conditions cannot yet be dismissed without further experimental work.

The belief that mycorrhizas are absent from some soils with a consequent failure in establishment, especially of forest trees, is supported only by a limited number of afforestation experiments in areas remote from trees of a similar kind. The demonstration by Robertson (1954) that propagules of mycorrhizal fungi were air-borne and ubiquitous in (as well as outwith) afforested areas so that seedlings became infected easily and routinely suggests that it would be unusual to find any newly afforested areas – except in remote locations – in which the trees had escaped infection either *in situ* or in the forest nursery before transplantation. There is, however, a great deal still to learn about the relationship between plants and their fungal partners and the soil. Modern concerns about depleting world resources of phosphate and the need for careful use of nitrogen to avoid pollution, together with a search for minimal-input systems on economic grounds, suggests that mycorrhizas will be of growing interest to foresters and agriculturists of the future.

The legume–rhizobium symbiosis

The nodules associated with the roots of legumes are very important in soil fertility but under the economic climate pertaining for most of our fifty-year period the role of legumes in soil fertility has been greatly neglected. The general connection between nodulation of the roots of legumes and the provision of nitrogen was well understood but the detail was not. A great outpouring of scientific work on the rhizobia and their part in nitrogen fixation took place in the 1940s and 1950s, well summarised by Nutman (1963). The science alone was not enough. It is doubtful whether its impetus would have been sustained without a series of practical advances in grassland husbandry to nourish it. The experiments on pasture improvement at Cockle Park in Northumberland in the 1920s and 1930s (Pawson, 1960) had demonstrated how clover fed with basic slag (which contained both phosphorus and lime) could contribute to the improvement of poor clay pastures. Stapledon, following his work in the 1914–18 war, succeeded in establishing the Welsh Plant Breeding Station and from it a range of valuable selections of grasses and clovers became available to farmers. The campaign to encourage the ploughing up of grassland during and after the war years gave an impetus to the use of these new clovers and grasses.

There was no doubt about the value of legumes as a source of nitrogen but after the Second World War fertiliser nitrogen became cheaper and the management problems of fertilising permanent or semi-permanent grass were less than those of the frequent establishment of new swards. The change in prices for fertiliser nitrogen, the problems of pollution from applied nitrogen in artificial fertilisers and the general move to reduce industrial inputs to a shrinking agriculture may bring a return of interest in clover. Clover and lucerne in temperate climates are joined by Berseem clover as a winter annual in semi-tropical countries; subterranean clover is of enormous importance in grazing situations in Australia. There is a particularly large number of leguminous species in tropical countries but their nitrogen-fixing capabilities have not been confirmed in all cases. Donald (1960) reviews the evidence and suggests, on a world scale, an annual production (mainly from symbiotic sources) of one hundred million tonnes of nitrogen. Detailed studies on the biochemistry of nitrogen fixation in the nodule have demonstrated the presence of haemoglobin, which plays an important part in oxygen transfer, and a biochemistry that is very dependent on the presence of a range of nutrient trace elements (molybdenum, boron, zinc, copper and cobalt) and the major elements phosphorus and calcium. Deficiency of one or more of these elements can explain failure to nodulate in a number of hitherto inexplicable cases. Other reasons for failure to nodulate have been shown by Nutman to lie in the absence of compatible strains of the organism and plant.

Where legumes like clover are plentiful inoculation of the seed is not likely to be necessary but where, as in lucerne, the crop may be a new introduction to an area, it is necessary to inoculate with a compatible strain of the bacterium. Raymond (1981) states that under optimum management a hectare of clover–ryegrass sward is likely to fix annually about 200 kg ha^{-1} of N, with a yield potential of about 7500 kg of dry matter compared with a potential yield of over 11 000 kg of dry matter from 400 kg ha^{-1} of applied nitrogen fertiliser. In New Zealand, where there is an all-the-year-round grazing regime and a much longer growing season for clover–ryegrass swards, the annual dry matter production will be much larger. In Britain the value of the clover clearly hinges on the value obtained for the 3500 kg dry matter (11 000 kg − 7500 kg) relative to the cost of the additional fertiliser. Other important factors are the skill and the objectives of management. The maintenance of a clover–ryegrass sward depends on skilled management and on maintaining a close-cropped sward; this maintenance regime may not be consistent with alternative objectives such as the use of the grass crop for hay or silage. If nitrogen fertilisers remain cheap enough then the greatest flexibility and perhaps the most economical production are obtained by their use. Where grazing is an important element in the enterprise the ability of stock to use the increased tonnage of grass obtained from fertiliser without trampling and spoiling it becomes a factor in decision making. It is of course possible to design a farming system that uses the nitrogen from clover–ryegrass swards as the main input and to top it up with fertiliser to increase the total production. It has been shown that the apparently adverse effects of nitrogen on clover are direct only in part. For the rest they stem from increased competition between the clover and ryegrass for light and water. However, such a system requires very high skills indeed in grazing management, and large resources in animals, for success.

Fertilisers

Four elements are commonly used in fertilisers: calcium, nitrogen, phosphorus and potassium. Calcium is applied as lime; it has been an objective of individual farmers as well as a policy objective for Government, using the scientific information outlined earlier in this chapter, to ensure that the soil pH was raised sufficiently to avoid crop failure due to soil acidity. This is achieved by regular inputs of lime in a variety of forms, particularly ground limestone. A subsidy on liming was paid to farmers in the period from 1937 to 1976, successfully seeking to encourage adequate applications of lime country-wide. In England and Wales lime was applied over the period at about 3 million tonnes of calcium carbonate per annum and it is reckoned that the annual requirement for the future is at that level. The campaign for increasing (1938–1950) and for maintaining (1950 onwards) lime status has been highly successful and there is ample evidence that the improvement in lime status of soils has increased yields in appropriate crops (Russell, 1973); however, the effects of the removal of the subsidy, combined with the recent drop in farming incomes, is not yet clear although there is some evidence that on the most vulnerable soils to the north and west of the country levels of liming are falling below the optimum (R. Speirs, personal communication).

The other three elements, nitrogen, phosphorus and potassium, have been applied to British soils over a very long period in the form of farmyard manure, imported guano (in the 1800s), shoddy, basic slag, town waste, bones, etc. The use of specific chemical inputs in the form of ammonium sulphate, superphosphate and potassium sulphate or chloride was well established by the beginning of our period (1936) but they were used at modest levels and were difficult to mix and apply because of deliquescence and binding of the mixture. They continued as the major fertilisers applied routinely in British farming until the early 1960s, either singly or increasingly as compounds in granular form, which allowed uniform distribution. The nitrogen and phosphorus components are now supplied almost entirely in the form of urea, ammonium nitrate and ammonium phosphates. The introduction of these substances made it possible to formulate fertilisers that were almost three times as concentrated, such that the wartime fertiliser with (e.g.) 7%N; 7%P_2O_5; 7%K_2O (7 : 7 : 7) has been superseded by fertilisers rated as 15 : 7 : 18; 23 : 5 : 9; 9 : 11 : 21, depending on their use. An almost incidental consequence of this was the virtual elimination of calcium, magnesium and sulphur from standard fertilisers.

At the beginning of our period, fertiliser application rates were at first derived from local experimentation, often carried out by the adviser and giving him first-hand information to pass on to his client. This was an expensive method, in terms of the use of the adviser's time. Under the stress of wartime, and using sometimes inadequate data, a comprehensive series of fertiliser recipes for various crops was produced. Extra supplements of phosphorus and potassium were allocated on the basis of soil analysis. In 1941, E.M. Crowther and F. Yates at Rothamsted analysed the results of a large number of field experiments and demonstrated the form of the response to increments of fertiliser under British conditions. Following Crowther and Yates, more sophisticated analyses were undertaken including mathematical models developed by D.J. Greenwood and his colleagues at the National Vegetable Research Station, Wellesbourne, for the prediction of fertiliser requirements of vegetables. Further work led to the production of an NPK Predictor for use by the advisory services. Such a bold initiative, although an advance on previous methods, has inevitable inaccuracies because of our inability to predict the weather and the expected supply of nitrogen to the crop directly from the soil reserves. This understanding of the soil's needs was accompanied by great advances on the part of the chemical manufacturers in the formulation of their products, which are now presented in granular form, easy to handle and slow to deteriorate in reasonable storage conditions, and as prills, which are coated chemical fertilisers (e.g. Nitram) produced by a very elegant and sophisticated manufacturing process. Advance in chemical formulation has been accompanied by the development of machinery for their application. Efforts have also been made to develop the use of liquid fertilisers and even to inject gaseous ammonia into the soil using a special machine. These methods bring logistic and economic problems and their success depends on particular circumstances.

Throughout the whole of the fifty years of the modern agricultural revolution, increasing knowledge of soil fertiliser requirements and the confidence that goes with it led to a very large increase in the use of fertilisers in British agriculture and a parallel increase in production from the land. Cooke (1981) demonstrates a 20-fold increase in nitrogen usage, a 2.5-fold increase for phosphorus and a 6-fold increase for potassium over an approximate 50 year period. Interestingly, although temporary grass crops have kept pace with arable crops in fertiliser usage the permanent grass has not, reflecting the economic and ecological constraints on production in some if not most of the areas of permanent grass. There is little doubt that the increased use of fertilisers contributed greatly to the increased farming output during the Second World War (Murray, 1955). Fertilisers were subject to price control from 1940 until 1951, and subsidies were provided on nitrogen and phosphorus until 1974. Nowadays arable crops mostly receive as much fertiliser as is required for optimum yield (and sometimes more), and applications to grassland have been making steady progress.

Nitrogen is the element to which crops respond most obviously. Arable crops such as barley and wheat respond to nitrogen up to about 150–200 kg ha^{-1} of fertiliser applied, depending on the nature of the soil, climate and variety. At higher applications the response flattens and indeed may be negative. Responses to this concentration of fertiliser had to await the breeding of short, stiff-strawed cereal varieties which were also physiologically capable of responding with increased yield. One of the problems of nitrogen fertiliser is that much of it remains in the soil solution and in autumn-sown crops can be leached from the soil in the autumn rains, necessitating compensatory applications later. When yield is plotted against nitrogen applied, the curve obtained is commonly parabolic, and it is easy to calculate the level of fertiliser to produce maximum yields. The fact that later increments of fertiliser show smaller responses than earlier increments, however, means that there is a rate of fertiliser application *below* the maximum which will yield the most profitable crop. There are other constraints nowadays with the realisation that nitrate contamination of drinking water sources is a growing problem, although it is not yet entirely clear how much of the polluting nitrate in fact comes from direct nitrogen fertilisation of arable crops and how much from the mineralisation of crop residues or recently added bulky manures such as farmyard manure, or even from the accumulated fertility of former years (Addiscott *et al*, 1991).

Phosphorus fertiliser applications to British soils raise other problems. British soils contain little natural phosphate derived from indigenous rock minerals. Most have large reserves of phosphate which has been accumulated over years of fertilisation with exotic phosphate sources. Much of this phosphate is unavailable to the crop but, as explained above, some becomes available through anion exchange and some by microbial action and mineralisation. However, it remains very difficult to predict the behaviour of soil phosphate in any particular year and it is common practice to include phosphate in the formulated fertiliser that is drilled with the seed, so that it is readily available to the emerging roots and taken up on germination, allowing later requirements to come from soil releases. In other parts of the world, such as the African tropics and Australia, soils respond markedly to phosphate and, especially in Australia because of the rapidity of soil fixation, it is placement-drilled in a band beside the seed. There is evidence that responses to fertiliser phosphorus have declined in Great Britain during our fifty-year period. Partly because of this, partly because of increase in nitrogen use, and partly because of withdrawal of subsidy, the ratio of P to N in fertilisers has gone down substantially.

Many British soils contain an adequate store of potassium, but on some of the soils derived from poor-quality sandstones and from glaciofluvial sands and gravels, and on peat-derived soils, the response of crops to potassium may be great.

The elements such as sulphur and magnesium, which are requirements of plants, are not generally specifically included in fertiliser mixtures, although they may be there by serendipity (ammonium sulphate), but they are applied if a deficiency is diagnosed. Nor are the trace elements generally added to the fertiliser mixture except for reasons of possible commercial advantage. Specific deficiencies of trace elements are reliably diagnosed from visual symptom; where doubt or a new problem arises they may be diagnosed using spectrographic methods developed, among others, by R.A. Mitchell at the Macaulay Institute. For a time a variety of ingenious biological methods developed by Hewitt and Wallace at Long Ashton and by Roach at East Malling Research Station were also used.

Organic farming appeared as a remarkable phenomenon in the period we cover. The memory of 'high farming' in the nineteenth century, when high yields had been obtained by the use of very high levels of organic manuring, (mostly with farmyard manure), remained even when the use of chemical fertiliser had become general. The work of Albert Howard in India in developing the Mysore process for the use of organic waste, and the proselytising of Lady Eve Balfour, pre-date the beginning of a large-scale movement in Europe for the production of 'healthy food' by organic fertilisation alone. This involved the provision of nutrients from organic wastes only, a perfectly feasible if expensive undertaking under modern farming conditions, but only by accepting reduced yields or 'borrowing' fertility from elsewhere. It also requires the avoidance of the use of chemical sprays against pests and diseases; this has still to be shown to be feasible in all years. The difficulty is that there is as yet no evidence to suggest that such a system is superior in terms of yield, or safety on ingestion; the question of the 'quality' of the produce is more difficult to assess and therefore to dismiss. Our concern here is with the equivalence of organic waste and artificial chemical fertilisers. Much of the evidence on this before 1960 derived from the Rothamsted long-term field experiments and from other shorter-term experiments in other parts of the

U.K. None of these was able to show differences in yield at equivalent dressings of chemicals, either straight or as farm yard manure. More recent experiments at Rothamsted by Warren, Johnson and D'Arifat (1964) using modified experimental techniques, with modern varieties and adequate pest control, showed some superiority in yield of the organically grown crops. It is possible to argue that the differences can be accounted for by the sustained supply of nutrients provided by the organic material, by the improved moisture regime, by the increased ramification of the rootlets or by the physical presentation of the nutrients in a soil expanded by the incorporation of organic material. However, there is not much evidence on these matters or on the question of crop quality, which is highly subjective matter and difficult to quantify. Much experimentation therefore remains to be done to elucidate the actual effects of organic manures in the soil and on the plant.

Soil structure

As soils came to be better understood, it became clear that there were unexplained examples of poor crop performance on soils that were apparently fertile. Some of this was attributed to the formation of impermeable layers in the soil: not, as in a podsol, from deposited materials, but as compaction due to repeated cultivations or the passage of heavy vehicles (Soane, 1984; Greenland, 1977). Effects of drought on soils are also considered by Pereira (1978).

The weight of tractors, their ability to work in conditions where the soil is plastic, and the nature of the cultivators they power, have combined to bring about deterioration in structure in certain soils. The identification of this problem has allowed farmers on vulnerable soils to take avoiding action.

Water and soil

At appropriate levels of fertility, the productivity of the soil is determined by the amount of water it receives and holds. Permeable soils under cold, high-rainfall conditions may be saturated and unproductive; under warmer conditions they may support good crops. In nature there is every possible combination of soil type–rainfall–temperature interaction. In agriculture – from the time of the ancient Egyptians – lands unproductive because of low rainfall have had their water supply supplemented by irrigation. The success of irrigation has depended on advances in engineering applied to its problems. These range from the earliest counterbalance systems, used by the Egyptians to this day (as the shaduf), to raise water by the bucketful to supply perhaps a hectare of land; through the Persian wheel, an endless circle of buckets which might supply sufficient water for perhaps six hectares of land; and the Archimedes screw of varying capacity depending on the motive power. Quite different are the wonderful *quanat* systems of Iran and neighbouring countries in which tunnels are carried into the alluvial fans on mountainsides to tap the permanent spring systems fed from the snows. In modern times dams and barrages with systems of canals and pumps servicing thousands of hectares have been constructed. The tube wells which tap underground systems by means of boreholes and pumps can service the water needs of single villages. Irrigation is important in certain parts of Britain, where it greatly improves yields in years of reduced rainfall, but it is not absolutely necessary for the growth of crops as it is in large areas of the arid and semi-arid tropics and subtropics. The emphasis in this section is accordingly on the problems of arid lands.

Irrigation brings with it major problems in soil science. Mishandled, an irrigation system can die. It can die within 30–50 years because of silting up of the reservoirs that hold its headwaters; modern research on dam construction has reduced that risk. It can die because of deterioration of the soil under irrigation. Irrigation water derived from rivers and aquifers shows varying degrees of salinity, from so-called sweet water, but still containing salts, to strongly saline water. In an irrigation system some salts in the applied water are left behind at the surface after evaporation; the speed at which they accumulate is related to the initial concentration of salts in the water. The great advances in irrigation have come from the study of this phenomenon and the discovery of ways of obviating its worst effects. When water can be applied in sufficient quantity, techniques can be developed for ensuring that the surface layers are sufficiently washed to prevent the development of surface salinity. Unfortunately pressures to accommodate more farmers than the system can reasonably service sometimes result in farmers at the margins of the area having to make do with less water than would maintain a healthy soil. Other salinity problems arise when the water table is near the surface of the soil and upward movement of the water allows the development of a saline surface. This is counteracted by ensuring that the subsurface drainage system lowers the water table; when ditches or tile drains are not feasible then resort can be had to tube wells, which can pump the water up from the aquifer and lower the level of the water table. They generally discharge the saline water into rivers, where it is diluted by river water, but this may have consequences for irrigation schemes downstream, which then receive higher levels of salt in the river water they use routinely. Much of the fundamental knowledge of the behaviour of saline soils under agriculture comes from the United States Soil Salinity Laboratory, set up in 1954 (Allison, 1964).

In Britain there are no areas where irrigation is an absolute prerequisite for agriculture, but there are areas where supplementary irrigation ensures profitable production. For example south-east Scotland and north-east England suffer drought, conveniently defined as a soil water deficit (the difference between the crop's requirement for optimal growth and the summer rainfall) of more than 150 mm in the period April to September, in 1–3 years out of 10 and hardly require irrigation. South-east England has a similar water deficit in 5–8 years out of 10, drought being still more frequent near the coast (Simpson, 1983). Yields of crops will be increased in the drought years by irrigation but decisions on whether to install irrigation equipment will be influenced by calculations of the cost in terms of water and the amortisation of the capital cost of the equipment measured against the value of the potential extra crop. The availability of water is also a factor. Independence from the supply is often achieved by the installation of small reservoirs fed by water extracted from a local stream in the winter, when water is plentiful. Where this is not possible the water demands of a large urban population will make the provision of sufficient irrigation water difficult; the complex laws governing riparian ownership and rights make summer extraction from watercourses very difficult.

Irrigation in arid lands makes technical demands on the farmers using it. The efficient distribution and utilisation of the water at the farm level must be matched by agronomic skill. The need for cooperation, self-denial and an enlightened approach to the use of water over the whole command system leads to similar progress in the technical and marketing aspect and the development of a highly advanced agriculture (Maas and Anderson, 1978).

In the discussion of irrigation so far it has sometimes been difficult to demonstrate a role for science in the empirical advances of practical men. However, apart from work at the U.S. Saline Laboratory, mention should be made of the Soil Physics Department at Rothamsted Experiment Station where work, started under B.A. Keen in 1913 and advanced by Scott Blair, E.W. Russell, R.K. Schofield and H.W. Penman, up to the end of our fifty-year period added to knowledge of the physics of the soil and especially of the role of water in it. The calculation of irrigation requirements and the provision of maps and estimates of need based on meteorological data are all practical derivatives of the development of the Penman Equation, which relates various parameters such as evaporation, humidity, sensible and latent heat, radiation, surface properties of plants, and wind velocity to the movement of water through and around plants (Penman, 1948). In the hands of skilled plant physiologists this equation or its modifications (Monteith, 1959; Monteith and Szeicz, 1961) can be used to answer a large number of questions about the water relations of the growing crop.

Once established by theoretical analysis, the measurements by which a soil's performance can be judged are readily used to determine the value of particular irrigation practices in the efficient use of water. In this way E.W. Russell (1973) identifies three important aspects. First, weeds, or inadequate fertiliser supply, waste water. Second, frequent irrigation leads to greater water use, which equates in part with water loss, because of more opportunities for water loss by evaporation. This compares with heavier and less frequent applications, which save water and do not reduce the production by the crop. Third, there are critical periods in the growth of a crop when water shortage will have serious consequences and other periods when water shortage may be less serious, but crops need a certain threshold to give any yield at all. Jensen and Musick (1960) show that winter wheat given 20 cm water yielded 0.78 t ha^{-1}; given 58 cm, the yield rose to 3.4 t ha^{-1}.

Water enters the soil as rain or as applied water in irrigation. It leaves the soil by evaporation, transpiration and drainage. Of equal importance to irrigation, therefore, has been work carried out on soil structure and on the percolation of water through soil and its removal by drainage. Much of this depended on empirical findings, and drainage practices became traditionalised. The theoretical basis of our understanding of soil water movement about the middle of our fifty-year period is summarised by Childs (1969).

Attention has been focused here on irrigation, but equally important is the work being carried out world-wide on the growth of crops under rain-fed conditions where water is limiting. Splendid work is being carried out by the research services in India, Pakistan and Australia, locally in Africa, and by the internationally funded research stations at Aleppo, Syria (ICARDA), and Hyderabad in India (ICRISAT).

An overview

At the beginning of our revolutionary period the classification of soils was in its infancy; by the end of the modern revolution there was an agreed world-wide natural classification. Once communication about soils, between scientists, became possible on the basis of this classification, work could begin on the examination of the mineral constituents, the humus and the chemical products of weathering. Knowledge could build up which allowed a rational consideration of the value of fertilisers and manures, trace elements and microbiological constituents. At the same time the role of water in the soil and in the plant was analysed.

The Soil Survey in England and Wales and in Scotland was issuing soil maps of great precision and utility right up to the end of our fifty-year period (when they came under threat of withdrawal of funding); as they gained in confidence, they, and others building on their information, began to issue derivative maps which indicated the quality of agricultural land and the suitability of land for recreation, housing, irrigation, minimal cultivation, etc.

By the end of the revolutionary period the soil, previously a mystery, had become understood as a complex but plausible working structure. This understanding allowed sensible farming decisions to be taken and brought precision into the growing of plants; it also provided a sound basis for the findings of agronomic science. This was achieved by the selfless work and the inspiration of a relatively few scientists. Some of these are mentioned here but there are many others whose contribution was equally vital but who for one reason or another have not found their way onto these pages.

Selected references and further reading

Addiscott, T.M., Whitmore, A.P. and Powlson, D.S. (1991). *Farming, Fertilisers and the Nitrate Problem.* C.A.B. International, Wallingford, Oxon, U.K.

Allison, R.V., Bryan, O.C. and Hunter, J.H. (1927). The stimulation of plant response on the raw peat soils of the Florida Everglades through the use of copper sulphate and other chemicals. *Florida Agricultural Experiment Station Bulletin* No. 190.

Allison, L.E. (1964). Salinity in relation to agriculture. *Advances in Agronomy* **16**, 139–80.

Avery, B.W. (1980) *Soil Classification for England and Wales (Higher Categories).* Rothamsted Experiment Station, Harpenden, U.K.

Burges, A. (1958). *Microorganisms in the Soil.* Hutchison University Library, London.

Childs, E.C. (1969). *Introduction to the Physical Basis of Soil Water Phenomena.* Wiley, Chichester.

Cooke, G.W. (1981). Soils and fertilizers. In *Agricultural Research 1931-1981,* ed. G.W. Cooke, pp. 183–202. Agricultural Research Council, London.

Crowther, E.M. and Yates, F. (1941). Fertilizer policy in wartime: the fertilizer requirements of arable crops. *Empire Journal of Experimental Agriculture* **9**, 77–97.

Donald, C.M. (1960). The impact of cheap nitrogen. *Journal of the Australian Institute of Agricultural Science* **26**, 319.

Dyer, B. (1894). *Fertilisers and Feeding Stuffs, their Properties and Uses.* Crosby Lockwood and Son, London.

Garrett, S.D. (1956). *Biology of Root-infecting Fungi.* Cambridge University Press.

Garrett, S.D. (1981). *Soil Fungi and Soil Fertility,* 2nd Edition. Pergamon Press, Oxford and New York.

Gasser, J.K.R. (1964). Effects of 2-chloro-6-(trichlormethyl)-pyridine on the nitrification of ammonium sulphate and its recovery by rye. *Journal of Agricultural Science, Cambridge* **64**, 299–303.

Greenland, D.T. (1977). Soil damage by intensive arable cultivation: temporary or permanent? *Philosophical Transactions of the Royal Society, London* B **281**, 193–208.

Hall, A.D. (1905). *An Account of the Rothamsted Experiments.* John Murray, London.

Harley, J.L. (1959). *The Biology of Mycorrhiza.* Leonard Hill, London.

Harley, J.L. and Smith, S. (1983). *Mycorrhizal Symbiosis.* Academic Press, London and New York.

Holmes, J.C., Donald, A., Chapman, V., Goldberg, S.P. and Smith, K.A. (1980). The problem of poor barley growth on light soils prone to loose seedbeds. *Annual Report of the Edinburgh School of Agriculture.* East of Scotland College of Agriculture, West Mains Road, Edinburgh EH9 3JG, U.K.

Jensen, M.E. and Musick, J. (1960). The effect of irrigation treatments on evapotranspiration and production of sorghum and wheat in the Southern Great

Plains. *Proceedings of the Seventh International Congress of Soil Science (Madison) 1960*, **1**, 386–92.

Kubiena, W.L. (1950). *The Soils of Europe; Illustrated Diagnoses and Systematics.* (Conseja Superior de Investigaciones Cientificas.) Thomas Murby and Company, London.

Maas, A. and Anderson, R.L. (1978). *. . and the Desert Shall Rejoice. Conflict, Growth and Justice in Arid Environments.* Massachusetts Institute of Technology Press, Cambridge, Massachusetts, and London, England.

Monteith, J.L. (1959). The reflection of short wave radiation by vegetation. *Quarterly Journal of the Royal Meteorological Society* **85**, 386–92.

Monteith, J.L. and Szeicz, G. (1961). The radiation balance of bare soil and vegetation. *Quarterly Journal of the Royal Meteorological Society* **87**, 159–70.

Mosse, B. (1963). Vesicular-arbuscular mycorrhiza: an extreme form of fungal adaptation. In *Symbiotic Associations, 13th Symposium of the Society of General Microbiology*, ed P.S. Nutman, pp. 146–70. Published for the Society by Cambridge University Press.

Murray, K.A.H. (1955). *Agriculture during the Second World War. History of the Second World War. United Kingdom Series.* Her Majesty's Stationery Office, London.

Nutman, P.S. (1963). Factors influencing the balance of mutual advantage in legume symbiosis. In *Symbiotic Associations, 13th Symposium of the Society of General Microbiology*, ed P.S. Nutman, pp. 51–71. Published for the Society by Cambridge University Press.

Nye, P.H. and Tinker, P.B. (1977). *Solute Movement in the Soil-Root System.* Blackwell Scientific Publications, Oxford.

Parks, D. (1955). Experimental studies on the ecology of fungi in the soil. *Transactions of the British Mycological Society* **38**, 130–42.

Parks, D. (1957). Behaviour of soil fungi in the presence of bacterial antagonists. *Transactions of the British Mycological Society* **40**, 283–91.

Passioura, J.B. and Leeper, G.W. (1963). Soil compaction and manganese deficiency. *Nature, London* **200**, 29–30.

Pawson, H.C. (1960). *Cockle Park Farm; an Account of the Cockle Park Experiment Station from 1896-1956.* Oxford University Press, London and New York.

Penman, H.L. (1948). Natural evaporation from open water, bare soil and grass. *Proceedings of the Royal Society, London,* A **193**, 120–45.

Pereira, H.C. (1978). *Scientific Aspects of the 1975-76 Drought in England and Wales.* The Royal Society, London.

Raymond, W.F. (1981). Grassland research. In *Agricultural Research 1931-1981*, ed. G.W. Cooke, pp. 311–23. Agricultural Research Council, London.

Robertson, N.F. (1954). Studies on the mycorrhiza of *Pinus sylvestris. New Phytologist* **53**, 253–83.

Robinson, G.W. (1936). *Soils, their Origins, Constitution and Classification*, 2nd Edition. Thomas Murby and Company, London.

Russell, E.W. (1973). *Soil Conditions and Plant Growth*, 10th Edition. Longman, London and New York.

Simpson, K. (1983). *Soil.* Longman, London and New York.

Smith, K.A. (1983). A model of the extent of anaerobic zones in aggregated soils, and its potential application to estimates of denitrification. *Journal of Soil Science* **31**, 263–77.

Soane, B.D. (1984). The economic consequences of the incidence of soil compaction by vehicles. *AGENG 84* (Proceedings of a Conference of the National Institute of Agricultural Engineering of the United Kingdom, April 1984), extended abstract, p. 48.

Tinker, P.B. (1979). Uptake and consumption of nitrogen in relation to agronomic practice. In *Nitrogen Assimilation of Plants*, ed. E.J. Hewitt and C.V. Cutting, pp. 101–22. Academic Press, London.

Waksman, S.A. (1947). *Microbial Antagonisms and Antibiotic Substances*, 2nd Edition. The Commonwealth Fund, New York.

Wallace, T. (1943). *Diagnosis of Mineral Deficiencies in Plants by Visual Symptoms*. Her Majesty's Stationery Office, London.

Warington, K. (1923). The effect of boric acid and borax on the broad bean and certain other plants. *Annals of Botany* **37**, 629–72.

Warren, R.G., Johnson, A.E. and D'Arifat, J.M. (1964). *Annual Report Rothamsted Experiment Station for 1964*, p. 40. Rothamsted, Harpenden, U.K.

6 The control of weeds, pests and plant diseases

The human battle against organisms that destroy or compete with crops goes back to the beginning of farming history. In the past it has been mostly a losing battle. Today the enemy is largely contained; the techniques for control of weeds, pests and diseases of crops put into practice over the past fifty years have to a great extent determined the success of the latest agricultural revolution. Some of the substances used, derived from the success of synthetic organic chemistry, are highly effective but, wrongly used, may pose a threat to the natural environment; others depend on sophisticated biological understanding and need skill to be effective. In this chapter we hope to give a balanced view of the development of these powerful tools of husbandry and to evaluate their importance.

Weeds

Until 1940 control of weeds was largely by rotation, by cultivation and by the production of crop seed free from weed contamination. The whole arable cycle was determined more or less by the need to control weeds. Indeed a farmer's skill would be judged by the success of his weed control methods as shown by the appearance of his farm. There is no doubt that farmers successful in weed control were generally successful in financial terms, but there was no absolute relationship and it was possible to go bankrupt with a clean and well-cultivated farm when control of weeds took precedence over careful management of marketing and finance. Estimates of the numbers of weed seeds in the soil vary from 20 000 000 to more than 250 000 000 ha^{-1}. They are long-lived; even when there is a surge of germination, some seeds will remain dormant to be available for a later opportunity.

Where weed control is practised intensively there is a decline in the number of weed seeds in the soil over a period of years. Hence weed control becomes easier with time on a well-managed farm. Most arable weeds are naturally plants of open ground and have been selected over millennia for the capacity to germinate quickly and grow rapidly, a capacity which is also useful when they

89

are present in the open ground of newly sown agricultural fields. Adapted as they are to specific ecological conditions of soil and climate, individual weeds are attuned to the cultivation cycle of particular crops and, if not controlled, can germinate in such numbers and grow so quickly that they smother the crop in its early stages. When good husbandry prevents this they grow later in sufficient numbers to compete with the crop for nutrients, light and water. In experimental plantings made to determine the effect of weed density on the growth of crops, the results when plotted show a sigmoid curve with low weed densities having little effect on yield, increasing densities having ever more severe effects until at the highest densities crop yield stabilises at a very low level (Harper, 1983). There is also evidence for a critical period for weed competition; in onions the bed can be left weedy for four weeks without apparent harm but from then on there is a rapid decline in the yield of the unweeded crop (Hewson and Roberts, 1971).

Sanders' (1943) textbook of crop husbandry epitomises the situation at the beginning of the modern agricultural revolution. A great deal of space is devoted to ploughing, fallowing, cultivating, hoeing, harrowing and manipu-lating the crop to achieve freedom from weeds. Not only were mechanical methods of hoeing and cultivating used but also gangs of workers were employed for hand-hoeing routinely and in emergencies. In the rich Fenland soils seedlings of onions and celery would be weeded by hand by workers moving along the rows on their hands and knees. In small plantation or subsistence agriculture in tropical countries, constant cultivation of the soil with the mattock ensured freedom from weeds.

Statutory control

The need to take account of weed contamination of crop seeds was recognised by the setting up of seed-testing stations in Scotland and England in 1914 and 1917, respectively. They were given statutory powers by the Seeds Act of 1920. A more positive approach to seed health and purity came in 1947 with the establishment of a Cereal Field Approval Scheme by the National Institute of Agricultural Botany.

Early chemical control

Chemical means of controlling weeds were also used. Charlock (*Sinapis arvensis*) in cereals could be controlled by spraying with copper sulphate, a method of weed control discovered in France when copper salts were being introduced at the end of the nineteenth century to deter pilfering in the vineyards and as fungicides against vine mildew and potato blight diseases. Accidental spillage of copper sulphate demonstrated its power of killing broadleaved weeds and it eventually became recognised as a useful pre-emergence herbicide and for use after emergence in cereals and waxy-leaved plants such as onions. Sulphuric acid followed for use in similar situations at a dosage rate of 5–12%. Its corrosive nature required specialised technical help and specialised machinery. The main problem with both materials was the large quantity of water required to obtain adequate coverage of the crop at the dilutions used.

The new weedkillers

The development of an industrial capacity to produce organic chemicals on a large scale, which took place in the first half of the twentieth century, saw the development, in France in 1932 of dinitro-orthocresol (DNOC) and later, in California, of 2,4-dinitro-6-secbutylphenol (dinoseb). Both could be sprayed on weed-infested cereals, killing the weeds but only slightly affecting the cereal. They could be used at high concentrations, but dilution reduced the risk of scorching. Their advantage lay in the absence of machinery corrosion; their disadvantage was that wrongly handled they could greatly harm the crop and their mammalian toxicity was high. Consequently they gradually fell out of use as other weedkillers were developed. They continued to be used, however, to reduce the cane vigour in raspberry crops in Scotland until the 1980s, when they were finally banned on account of their possible toxicity to human operatives.

The next development took place during the early years of the Second World War and demonstrates a common phenomenon in science where two groups of workers simultaneously make the same discovery, sometimes, as here, by entirely different routes, and in ignorance of one another. A group working for Imperial Chemical Industries at Jealott's Hill were interested in the recently discovered plant hormones and in synthetic materials with similar properties. Work on plant hormones has a long pedigree. Probably the first work was that of Sachs, who was interested in the mechanism of growth tropisms, work continued by Darwin in 1890 when he demonstrated a positive phototropic curvature in the tip of the grass coleoptile, which thereafter became a favoured experimental tool. The coleoptile is the first 'leaf' to emerge as a delicate white tube-like sheath from the germinating seed; the first green leaf emerges from its apex. Excision of the coleoptile apex destroys its capacity to respond to light by bending towards it. There followed a whole series of discoveries that demonstrated the critical role of the coleoptile tip in producing the substances responsible for the bending. These substances were isolated by Went (1928) in agar plates into which they diffused. Similar substances were found in human urine by Kögl and Haagen Smit (1931); one of them was found to have the same growth-regulating properties as Went's material and was a known compound, β-indoleacetic acid (IAA). Its presence in plant tissues was confirmed and it was this material and its analogues that the ICI scientists at Jealott's Hill were investigating in a search for any long-lasting effects on plants. Could plants be made bigger or smaller by treating them with these substances? No giant or dwarf plants were discovered but at certain concentrations α-naphthylacetic acid and similar synthesised substances incorporated in the soil were seen to prevent the growth of plants, which showed an initial spurt of elongation and then assumed a twisted and spiralled appearance which became more severe until the plant died. Broadleaved plants reacted at lower concentrations, and more severely, than monocotyledonous plants (grasses and cereals). It was not difficult to build on these findings and demonstrate that broadleaved weeds could be killed in cereal crops (Templeman and Sexton, 1946). A wide range of related compounds was examined, culminating in the identification of 2-methyl-4-chlorophenoxyacetic acid (MCPA) as a cheap and effective post-emergence selective weedkiller.

At the same time workers at Rothamsted studying the invasion of clover roots by the nodule organism *Rhizobium leguminosarum* (Nutman, 1963) noted that the root hairs of clover became deformed at the point of invasion by the nodule organism, following conversion of secreted tryptophan to indoleacetic acid. Further examination of this substance and naphthalene acetic acid demonstrated that they were readily degraded in contact with soil; an examination of the degradability of similar compounds eventually demonstrated that one of these, 2,4-dichlorophenoxyacetic acid (2,4-D), was resistant to breakdown in the soil and was also a highly effective post-emergence selective weedkiller.

Because of wartime secrecy the results from Jealott's Hill and Rothamsted were not immediately published but were communicated to the Agricultural Research Council, which brought both sets of workers together with Professor Geoffrey Blackman, then Head of the A.R.C.'s Unit of Experimental Agronomy at the Department of Agriculture of Oxford University. The outcome was the development of a whole family of post-emergence selective weedkillers, some from parallel work in the U.S.A., which then appeared as miraculous compounds and which are still in use throughout the world. The detailed history of these developments is well described by Peacock (1978).

There was now sufficient evidence of the relationship between chemical structure and biological function in auxin-derived herbicides for R.L. Wain, at Wye College, to begin a systematic study of their chemistry, synthesis and activity. This work was recognised by the Agricultural Research Council's foundation there of the Unit of Plant Growth Substances and Systemic Fungicides under Wain's direction. One outcome of his work was the recognition that plants varied in their capacity to metabolise fatty acids, so that substituting butyric acid for the acetic acid in MCPA and 2,4-D offered immediate protection to cereals, clover, lucerne and certain vegetable crops while a range of noxious weeds used their β-oxidase enzyme to convert the relatively harmless butyric acid radical into the poisonous acetic acid, thus committing suicide. It is no exaggeration to say that the development of these and other selective weedkillers revolutionised arable farming from the 1950s forward.

A remarkable philanthropic spin-off from this period was the foundation by Louis Wain of the Wain Fund, based on royalties from the new herbicides discovered by the Unit. This fund, held in trust by the Agricultural Research Council, is used to provide travel grants and fellowships for younger research workers.

Once the commercial value of herbicides was understood, this intellectually satisfying approach to the identification of new herbicides gave way to the systematic testing and evaluation of available chemicals with little concern for their structure or biological function. Mundane, perhaps, but very rewarding these methods were. All the major chemical companies of the world took part. The range of compounds discovered, their uses and methods of application can be found in the *Weed Control Handbook* produced in several editions by the British Crop Protection Council under the initial editorship of staff of the Weed Research Organisation, which grew out of the A.R.C.'s Unit of Experimental Agronomy. The types of weedkillers that emerged can be classified as follows, always remembering that some compounds have characteristics that bring them into two of the categories in each group.

A sample of modern weedkillers

They are classified first on whether they are used in the crop:

pre-sowing (or pre-planting);
pre-emergence;
post-emergence.

Within each category they may be:

contact herbicides;
translocated herbicides;
residual herbicides.

A selection of the compounds found includes the following.

Dalapon. A halogenated aliphatic acid (Dow, U.S.A., 1956) which could kill grass weeds such as couch grass. A pre-sowing translocated herbicide.

Diquat and paraquat. Bipyridyls (I.C.I., U.K., 1958, 1960) which could kill most annual weeds on contact. Pre-sowing or pre-emergence contact weedkillers.

Atrazine and simazine. Triazines (developed by Geigy in Switzerland in 1957) which could inhibit weed seeds but allow specific crop seeds to germinate and establish. They act as pre-emergence, selective, residual weedkillers.

Propham, carbam, asulam. Carbamates (Spencer, U.S.A., 1941; Fisons, U.K., 1960). Propham is a post-emergence, probably contact selective weedkiller used among others against a wide range of weeds in beet. Asulam is a post-emergence translocated selective weedkiller used against *Rumex* in established pasture and weed grasses in pasture.

Monuron, diuron, linuron. Urea and uracil derivatives (Dupont, U.S.A., 1957–60) which act as long-lasting residual pre- and post-emergence weedkillers for use on perennial crops at high concentrations and which may show selectivity at lower concentrations.

Ioxynil. Benzonitrile (developed simultaneously by May and Baker, U.K., and AmChem, U.S.A., 1964), a post-emergence contact weedkiller which has the valuable attribute of widening the spectrum of weeds killed by other post-emergence weedkillers.

Dicamba and mecoprop. Substituted benzoic acids (Velsicol, U.S.A., 1961; Boots, U.K., 1956) which are broad-spectrum post-emergence translocated weedkillers, extending the range of the MCPA type weedkillers, and can be used *ad hoc* in special situations, particularly the control of woody weeds.

Glyphosate (Roundup). N-(phosphonomethyl)glycine. This is one of the most remarkable of the recent discoveries. It was discovered in 1971 by J.E. Franz (Grossbard and Atkinson, 1985). It is a broad-spectrum, non-selective, post-emergence, translocated weedkiller for herbaceous perennial weed control, with high unit activity. It appears to work by inhibiting the enzymes of the shikimic acid pathway for the synthesis of aromatic amino acids, important for protein synthesis. It is held tightly to soil particles and rapidly broken down by soil microorganisms. In the plant it is rapidly transported to the meristems and after some time is metabolised. It is inherently of low toxicity because of the absence of the shikimic acid pathway in animals.

Those were the weedkillers of our period; they were succeeded by others such as the aryl compounds decoflop-methyl, an auxin inhibitor, the sulphonyl ureas and the imidazolinones (Kirkwood, 1991). With these and many other weedkillers a complex technology has grown to encompass dosage, mixture and application. At the same time it has been important to establish the mammalian toxicology and the persistence in the soil of all weedkillers. Most of those mentioned above have a low mammalian toxicity, but care has to be taken in spraying to avoid contact with the human operative because of the dangers of skin irritations and as yet unknown effects on health. Paraquat, although of moderate toxicity at spray dosage, is frequently fatal when ingested at high concentration; 2,4,5-T (2,4,5-trichloroacetic acid), a brushwood killer, has been suspected of teratogenic properties, although the evidence for this is in dispute (Conway and Pretty, 1991). The discovery of these substances not only changed the normal cycle of weed control operations but also was responsible for change in spraying techniques in Britain. The possibility of using selective weedkillers at low volumes led first to the adoption of American-designed booms and nozzles, spraying at rates as low as 2–8 gallons acre^{-1} (20–90 l ha^{-1}), and then to improved local design.

The introduction of paraquat and later glyphosate opened the eyes of agronomists to possible new systems of cultivation such as direct drilling and minimal cultivation (Chapter 8). The first of these allows the direct re-seeding of killed pasture for subsequent renovation, or the direct drilling of cereals and kale into killed grassland or stubble using a specially designed slit-seeder. The next step was to combine weed control by these weedkillers with minimal cultivation for both cereals and kale (Peacock, 1978). Indeed, the development of this method, where kale is broadcast onto killed grassland and then lightly stirred with specially designed rotary machines, is now a standard method for establishing catch crops for sheep feeding over much of the upland farming area in Britain. In addition, these weedkillers have been used more recently in 'no-till' farming situations in (for example) the United States to reduce soil erosion and moisture loss. An increase in their use can be related to anti-erosion legislation in the U.S.A. and to soil erosion problems in the rest of the world.

The production of such an armoury of herbicides may have revolutionised the work of the farmer and at the same time, as will be argued elsewhere, increased and cheapened the production of the world's staples, but the release of large quantities of highly reactive chemicals into the environment has not been without effect. The Ministry of Agriculture in the U.K. began to monitor the presence of these and other potentially hazardous herbicide chemicals in the ground water in the 1980s (*The Effects of Pesticides on Human Health*, Report and Proceedings of the House of Commons Select Committee on Agriculture, vol. 1, H.M.S.O., London, 1987) and reported that atrazine, simazine, mecoprop and 2,4-D were then found at concentrations of 0.02 – 0.4 μg l^{-1} in chalk aquifers; these concentrations gave no cause for alarm in terms of the then accepted British toxicological standards. Nevertheless they regularly exceeded the 0.1 μg l^{-1} allowed by the E.C. Drinking Water Standards Directive and were finally banned for non-crop use on 31 August, 1991. Restrictions for crop use were also introduced at that time. Whether any level of contamination is acceptable is arguable.

The control of pests

Increasing awareness of the losses caused by pests has been paralleled by great advances in the methods available for their control. Pests include animals of all types, from elephants to insects, from rabbits to eelworms, in situations where human food or comfort or livelihood are threatened by the excessive multiplication of the pest species. However, the principal pests world-wide are the insects and arachnids.

Estimates of the number of species of insect varies but it is probably in excess of two million, only about half of which have been recognised and described. Arachnids have far fewer species, but the mites are highly developed and important pests of plants. Potential pests generally have great powers of multiplication and can develop from a small infestation to overwhelm the crop in a short time, given suitable conditions. Even the elephant cited above can become too numerous for human comfort given enough time. Some pests live on a wide variety of plants in nature and multiply to threaten crop plants locally or (as in the case of the locust) across continents. Others depend on their host plant for breeding sites and sustenance and in seeking these may multiply to devastate the host (e.g. *Psila rosae*, the carrot fly) or to weaken it so severely that it becomes uneconomic (e.g. Pine beauty moth, *Panolis flammea*, on lodgepole pine). Others, such as the apple codling moth (*Cydia pomonella*) and plum moth (*C. funebrana*) are placed, as eggs, in or near the developing fruit, which the larvae enter to emerge as the fruit ripens, leaving evidence of their presence and causing cosmetic damage. The arachnids appear as mite attacks, commonly as 'red spider' infestations of the leaves, which severely weaken the plant and reduce the yield.

In animals there are a number of dipteran flies that lay eggs in the skin of cattle (warble fly, *Hypoderma lineatum*) causing discomfort and economic loss; or in sheep (the greenbottle flies, *Lucilia* spp.) near wounds or sores, where the hatched larvae eat their way into the skin and can cause large and sometimes fatal wounds. The arachnids attack domestic animals in two main ways, as skin infestations (mange) caused by small mites and then as blood-sucking large ticks which, apart from the debilitating effects of a heavy infestation, also transmit diseases such as louping ill of sheep and grouse and Lyme disease of deer, occasionally transmissible also to man and other mammals.

The immediate human response to pest insects or arachnids in epizootic proportions is to find some means of destroying them, possibly at a vulnerable point in the life cycle. For this an insecticide is needed; until the 1940s the range of material available was limited. In domestic situations, e.g. in glasshouses, the plant products nicotine and quassia were used to destroy aphids and, with some mechanical effort, to control scale insects. Derris and pyrethrum were recognised as having a wide range of uses before the Second

World War and on a field scale arsenical washes were routinely applied, for example in apple orchards. Nicotine was toxic to man, quassia of doubtful effectiveness, derris and pyrethrum difficult to formulate as the natural product (and in any case they were unstable in sunlight), and the arsenicals were poisonous to a wide range of animals and used on fruit at high concentrations over a number of years could effectively sterilise the orchard floor.

Synthetic pesticides

For the same reasons that led to the development of synthetic herbicides, the search for synthetic insecticides began in the 1930s, resulting in the discovery of lindane (the γ-isomer of hexachlorocyclohexane) in 1936, the organophosphates in 1937, DDT (dichlorodiphenyltrichloroethane) in 1939, the polycyclic chlorohydrocarbons in 1944, the carbamates in 1948 and the industrial pyrethroids in 1949. Each of these groups has seen its uses extended and in some cases the development of a range of cognate compounds over the next 30–40 years.

They were powerful substances. The first organochlorine insecticide, lindane, was developed by Imperial Chemical Industries Ltd and others and was quickly followed by DDT, discovered by Paul Müller, working for the Swiss firm of J.R. Geigy, in 1939. This discovery was the culmination of a long series of syntheses aimed at the production of an insecticide that would meet a set of criteria Müller had set for himself: rapidly toxic to insects, relatively harmless to warm-blooded animals and plants, non-irritant and virtually odourless, wide application, stable and long-lasting, cheap and easy to produce (Mellanby, 1989). DDT was truly a wonder compound; Mellanby has written (1992) about its role in the preservation of the health of troops in the Second World War by controlling insect-borne human diseases. Its use on water surfaces and as spray application in light petroleum oils to the walls of houses in tropical areas protected against the anopheline mosquitoes which carried malaria and, depending on the geographical area, yellow fever and a number of other important diseases. It also controlled the human body louse, which in earlier wars had spread the rickettsiae of epidemic typhus fever. Its use soon spread to agriculture, where it controlled insect infestations of animals and with equal ease could be used in orchards to protect against a variety of pests previously controlled by arsenical washes. It was equally useful against the insect pests of field crops such as carrot and turnip fly and the leatherjackets and wireworms. There was an early realisation that it was so powerful that its use on a field scale might have deleterious ecological effects by destroying a wide range of insects and permanently impoverishing the insect fauna of the countryside. Lindane and DDT were followed by aldrin and dieldrin, equally potent and persistent, used as seed dressings to control wheat bulb fly, as a sheep dip, and in the protection of wool-based fabrics against moth attack, as well as in the treatment of insect-induced deterioration of the timber in buildings.

Environmental concerns

Suspicions about the environmental effects of these compounds found a voice in *Silent Spring*, a book by Rachel Carson (1963). This articulated in a popular fashion the findings of environmental scientists in most of the developed countries, but particularly in the U.S.A. and United Kingdom, where discoveries were being made of specific instances of organochlorine-related environmental disaster. In the U.K. the use of seed dressings of wheat, in itself designed to minimise the amount of toxic material applied to the crop, led to ingestion of organochlorines by seed-eating birds, some of which died, while others survived but accumulated the materials in their tissues. When they were eaten by predators such as sparrowhawks and peregrine falcons, it was suggested that the decline in fertility in the falcons could be related to the ingestion and accumulation of organochlorine chemicals. One of the most convincing pieces of evidence was the demonstration by Ratcliffe (1970) of the relationship between the decrease in shell strength in the eggs of raptorial birds and the concentration of organochlorines in their tissues. There is also no doubt that the dramatic decline in the population of raptorial birds was encouragingly reversed when the use of organochlorine insecticides was reduced following legislation for their control. Similar serious effects have been noted in the bat population. Bats are particularly sensitive to organochlorines and apparently accumulate the materials in their tissues from the insects they eat. These materials are believed to be present in the body fat, where they are relatively innocuous, when the bat goes into hibernation; and as the bat comes out of hibernation the fat has been used but the organochlorines are now present in a more available form. The spraying of roof spaces in bat-infested buildings is another hazard for them. But not all the decline in the bat population can be placed at the door of the organochlorines because the increased human activity in the countryside puts at risk the preferred habitats of the bat population and the reduction in insect life from a wide range of recent farming practices also threatens their food supply.

Organophosphates and carbamates

Another important group of insecticides, the organophosphates, was discovered by G. Schrader at the Bayer laboratories in 1937, following chemical synthesis accomplished earlier by Lange in Berlin in 1932. The toxicity of these compounds was soon established and much of the chemical research carried out on them in the early days of World War II was directed towards military use (Naumann, 1989). The potential of those that showed less toxicity towards human beings was also explored, first by Bayer, then by I.C.I. and then, as their potential became appreciated, by many other chemical manufacturers. From this work came 147 separate organophosphate compounds developed as insecticides. Much praise must go to the skilful synthetic chemistry that produced them. Of this large number of compounds, many fell by the way for one reason or another until in 1986, according to Naumann, there remained some 18 well proven insecticides, which are responsible for a great deal of the routine insecticidal treatment in the world; there are currently 32 organophosphate compounds registered as insecticides in the United Kingdom. Familiar names among them include: chlorpyrifos, fenitrothion, dimethoate, malathion and fonophos.

The organophosphates were followed by the carbamates, the first of which was discovered in 1948; in all 45 had been discovered and developed by 1989. A much smaller number remained as proven insecticides by 1989 including such familiar names such as carbofuran, carbaryl, aldicarb and propoxur. For an outline of the chemistry of these substances and of the organophosphates, reference should be made in the first place to Naumann (1989).

Both the organophosphates and the carbamates include materials with a range of qualities: a wide or narrow spectrum of species kill; selective attack on nematodes, mites, or insects; water-soluble or not; attack by contact or by the stomach route; persistence for short or long periods; some act as a fumigant gas. A most valuable and novel quality of some of these insecticides is their systemic action in the host plant. After absorption they are available and lethal to sucking insects and may be translocated into freshly developing leaves, thus affording a degree of latitude to the sprayer and an insurance against any hidden sucking insects or those with a particularly resistant cuticle.

Although these insecticides have more than fulfilled expectations and are responsible for a large part of the market in insecticides world-wide, they are not without their problems. The pests they control may develop resistance, a major problem with insecticides in general. From the operative's point of view, although their mammalian toxicity is subject to stringent tests before release, no tests can legislate for accidental exposure to large doses, or determine possible long-term effects on those who come in contact with them at lower doses over a longer time span, or the dangers of sensitisation to the compounds. Cases of poisoning from exposure to these substances are relatively common on a world basis. The symptoms are those associated with acetylcholinesterase interference: headaches, giddiness, blurred vision, tachycardia and convulsions. However, it is only in extreme cases that they are fatal. More difficult to evaluate are the possible long-term effects as carcinogens or as the cause of delayed neuropathies with tingling and burning of the hands and feet, weakness and paralysis. Even more difficult are the reports of behavioural changes such as impairment of memory or speech and depression, irritability or anxiety. Sufficient doubt is present, however, to make the search for other insecticides worthwhile (Conway and Pretty, 1991).

The Rothamsted contribution

Although other complex organic chemicals have been synthesised and developed as insecticides, none has shown a capacity to compete with the organophosphates and carbamates except the synthetic pyrethroids. These have an interesting history. Whereas the origins of the organophosphates and the carbamates lay in innovative chemical synthesis, followed by testing for appropriate insecticidal properties, the pyrethroids began life as pyrethrin, the natural product of *Chrysanthemum cinerariaefolium* and other *Chrysanthemum* spp., extracts of which had been used in the Middle East for some hundreds of years. The story in Britain commences with the appointment of Frederick Tattersfield, a chemist, to the staff of Rothamsted Experiment Station in 1919, to begin the study of insecticides. After early studies of the relationship between structure and toxicity, ranging widely over the chemical field, and incidentally uncovering chloropicrin (a soil sterilant) and 3,5-dinitro-orthocresol, he turned his attention to the natural fish poisons, some of which were known to have insecticidal properties. Studies on *Derris elliptica*, a Malaysian plant, of which the powdered root has fish-poisoning properties, led to the isolation of the poisonous principle, rotenone. Work followed on *Tephrosia* spp. and *Lonchocarpus* spp. but, for the reason that it was an easily grown field crop, finally settled on pyrethrum (*Chrysanthemum cinerariaefolium*). Work at Rothamsted determined the optimum conditions for growing, harvesting and storing the pyrethrum flowers and led to the development of a considerable industry, at that time in the higher and cooler parts of Kenya. Tattersfield retired in 1946 to be followed by Charles Potter, whose work with a number of colleagues further unravelled the nature of insecticidal action (Russell, 1966). A renewed interest in pyrethrum led to the isolation of the insecticidal principles from the pyrethrum flower and, with the recruitment of Michael Elliot to the team, to the development of a range of synthetic pyrethroids, which were quickly developed by industry with a peak of patent applications in 1980. It is estimated that, in 1987, an area of 70 million hectares was treated with synthetic pyrethroids world-wide, and that the materials had an annual sales value of about £900 000 000 (Leakey, 1985; Elliot, 1989). So far the pyrethroids are unsurpassed for safety, high potency and selectivity; in addition, in comparison with the natural product, they are stable. Paradoxically, although some of the compounds are highly toxic to humans these same compounds are toxic to insects at such low concentrations that they pose little threat to the human operative. However, fish are less tolerant; accordingly, the effects of these substances in the aquatic environment need to be carefully monitored. Some are readily degraded, however, and this adds to their safety because whereas they are rapidly absorbed by insects and interfere with activities in the nerves of the insect (thought to be the sodium channels), on the human skin and on ingestion they are rapidly converted before they can reach sensitive sites of action. Some are finding uses as residual insecticides in the soil and show persistence under these conditions.

Biological control

The straightforward development of these three major classes of insecticides, the organochlorines, the organophosphates and the synthetic pyrethroids, broadly encompasses the contribution of pesticide chemistry to the second agricultural revolution. But while this work was producing immediately applicable results, other more exotic biological work was commencing. At first this was only a promise, but by the end of the revolutionary period some parts were beginning to develop into a workable methodology for more precise and user-friendly insect control. The oldest of these other aspects was biological control by the use of predatory insects, fungi, bacteria or viruses. Early efforts epitomised by, for example, the control of *Opuntia stricta* in Australia by the cactus moth (*Cactoblastis cactorum*) (Dodd, 1940) were often spectacularly successful but the inherent dangers of introducing exotic species into a new environment led to caution and the preference for control by the emerging new herbicides and insecticides. However, increased awareness of the hazards to the environment posed by these materials, the desire for species conservation, interest in the production of organically grown food, and concern about the emergence of resistance to herbicides, insecticides and fungicides, encouraged renewed interest in biological methods of control. In the United Kingdom biological control is currently most commonly found as part of the integrated control procedures developed by Hussey and others (Rudd Jones, 1981; Hussey, 1990) for the control of glasshouse pests. The parasitic wasp *Encarsia formosa*, introduced into glasshouses after a preliminary introduction of the whitefly pest *Trialeurodes vaporariorum*, is an effective control for the latter. *Phytoseiulus persimilis*, a Chilean predatory mite, parasitises the glasshouse red spider mite (*Tetranychus urticae*) and can also be used in a glasshouse situation. In the integrated method of control if the pest gets out of hand, i.e. if it multiplies faster than the predator, as happens at some time in most seasons, resort is had to spraying. In the two pests mentioned a complete kill is not likely (indeed is not sought); the pest and parasite recover together and a balanced population is restored.

The use of nematodes for the control of vine weevil is currently a possibility and the use of nematodes as a control of slugs is being promoted as a tool for amateur gardeners.

The use of fungi for the control of insects was tried out as early as 1877 and 1899 by Metchnikoff and Kassilschik (Charnley, 1989) on pests of wheat and sugar beet. Recently interest has been renewed, especially in the use of the fungi in an inundative fashion, where the environment is presented with a large volume of the inoculant. There are differences between such a use of an endemic parasite and the use of an exotic parasite in a new situation. Performance so far has been mixed but increasing understanding of parasitism and improvement in techniques may offer hope of greater success.

The use of bacteria and actinomycetes in the control of insects has been much more successful. *Bacillus thuringiensis* (Hannay, 1953) produces material within the spore (parasporal bodies). They appear as diamond-shaped crystals of mainly proteinaceous material, which, when ingested by insects, break down to produce toxic amino acids. These paralyse the gut and lead to death. The organism exists in a number of pathotypes that attack specific groups of insects. Successful attempts have been made to use the organism either alive or dead, or its extracted toxins, to control insects in specific situations. Jutsum *et al.* (1989) claim that *B. thuringiensis* derivatives now account for about 1% of the world market in pesticides. The actinomycete *Streptomyces avermitilis* produces macrocyclic lactone toxins which, although only recently discovered (Putter *et al.*, 1981), are widely used. They have acaricidal and nematicidal as well as insecticidal properties and it is the first two properties which, to date, have had particular success in the animal health field (Chapter 11).

Insect viruses are well established as a means of controlling insect pests (WHO/FAO, 1973), particularly the baculoviruses, which produce polyhedral crystals containing virus nucleic acid within the body envelope of the dead insect. These can be extracted and stored to contaminate and infect healthy insects, which in turn produce more virus, which is spread over the plant and affects any remaining healthy insects. These viruses have provided particularly useful control measures in forests where, for example, the spruce sawfly produces large populations of caterpillars, which are an easy target (Entwistle *et al.*, 1983). The viruses are very specific to particular insect hosts; thus control by virus is very safe for other organisms and, because the virus is ultimately degraded, the environment is not at risk.

Use of sterile males

There remains one important technique for insect control, apart from those that are still somewhat speculative and which will be proved in the future. It is the use of genetically sterile insects to reduce the population. The technique involves the release of sterile male insects, which, by mating with normal females in the wild, reduce the population of viable offspring to very low levels. The method has been used against the tsetse fly, the Mediterranean fruit fly and the Japanese melon fly with success, but potentially the greatest success has been the attempt to eradicate the new world screw-worm (*Cochliomyia hominivorax*) from North Africa. This fly, the larvae of which attack and feed on the living flesh of warm-blooded animals, including man and his domestic animals, entered Libya from South America, probably in 1988. It spread quickly to infest an area of about 25 000 square kilometres around Tripoli, and had the capacity to spread into southern Europe, all of North Africa and the Middle East and Western Asia. A United Nations Food and Agriculture Organisation eradication campaign, based on experience in the U.S.A., was launched with the release of sterile males, produced in Mexico. The work began at the end of 1990. By May 1991, 40 million pupae per week were arriving in Libya and were distributed in a staggered fashion, by air. There was a dramatic fall in the number of infected cases and none were reported after October 1991, allowing the authorities to discontinue the campaign, apart from surveying for any recurrence. The campaign cost about £20 000 000, about one third of the expected and estimated cost. Most significantly, it was a magnificent vindication of the role of FAO and international cooperation in pest control (Rogers, 1992).

Other approaches to the control of insect pests which depend upon increased understanding of the physiological control of the insect life cycles, and through the pheromones that control aspects of the mating process, and the hormones that control the complex physiology of pupation and moulting, are already beginning to be used experimentally. These are likely to become important pest control methods in the future.

Rodent control

Insects and arachnids are not the only pests with which agriculture has to contend. Of all the rest of the animals that may at times be a nuisance to the farmer, the rodents, particularly rats and mice, and rabbits, are a continuing problem on the farm. At the beginning of our fifty-year period the rats and mice were always associated with the storage of unthreshed grain in the farmyard and, after threshing, in barns. Although local epizootics of these pests might occur, the dispersion of the harvesting activity and the large amount of human and feline activity in the farmyard kept the problem within bounds. The quantities of grain now harvested and the concentration of storage in large buildings means that, when ingress has been achieved by the rodents, local multiplication can present serious problems. A common control method is fumigation (in suitable buildings) by cyanide (hydrogen cyanide), phosphine (hydrogen phosphide), methyl bromide or carbon dioxide. All were developed for this purpose during the last fifty years; all are effective but inherently dangerous. The alternative to fumigation is poisoning. Over the years a number of acute poisons have been developed. They include α-chloralose (a sedative relative of chloral hydrate), calciferol (vitamin D), lindane (the same as the insecticide of the same name, used as a contact poison against mice), organofluorines, reserpine, thallium sulphate and zinc phosphide. These are useful in different situations in the hands of experts. However, they mostly show bait shyness: the rodent fails to return to the baited food because it recognises a taint or associates the food with malaise. Above all they are poisonous; some of them are very dangerous to domestic animals. Some of the fumigants and some of the poisons have been banned in particular countries.

Undoubtedly the most important advance was the discovery of anti-coagulant rodenticides in the early 1950s. Dicoumarin was in fact synthesised by Anschutz in 1903 but its functional qualities were not recognised at that time. Coumarins were then recognised as constituents of some clovers and grasses and as contributors to the smell of some new-mown hays. In the 1920s spoiled Sweet Clover (*Melilotus alba*) hay was associated with a haemorrhagic disease of cattle, but although interference with the prothrombin of the blood was postulated as the aetiologic factor, the connection with dicoumarin was not made. Eventually a farmer took a haemorrhaging cow to the Wisconsin State Experiment Station; the responsible factor was sought in the Sweet Clover hay and finally in 1939 an anti-coagulant dicoumarin close to Anschutz's synthetic dicoumarin in structure was isolated. The potential value of this substance with its anti-coagulant properties was recognised, and patents were taken out by the Wisconsin Alumni Foundation. A series of similar compounds was synthesised in an effort to find a faster-acting material. The forty-second compound tested (3-α-phenyl-β-acetylethyl)-4-hydroxycoumarin was the best and was named 'warfarin'. It went on to become a major rodenticide and was also used to control clotting in circulatory diseases. It was only toxic when ingested in sufficient quantity so that accidental ingestion of small quantities by domestic animals or human beings was not catastrophic nor, within limits, was the ingestion of warfarin-poisoned rodents by cats or dogs. It could, accordingly, be made freely available to rodents and because of its bland taste did not induce bait shyness. Moreover, although it could not be assumed that no pain was inflicted, death from warfarin poisoning was not associated with any violent physical symptoms or nervous agitation. The action of warfarin was cumulative and mice in the early stages of poisoning seemed to carry on as usual, only collapsing towards the end of the baiting period of a few days.

However, as with many highly successful pest control materials, rats and mice became resistant to the action of warfarin in discrete geographical areas. As a result a number of other coumarins were synthesised and released as substitutes for warfarin, e.g. difenacoum and brodifacoum. Not all were as easy to use and some might be more threatening to domestic animals, but they remained useful rodenticides, to be preferred to any of the acute poisons (Meehan, 1984).

Failure to control the rabbit

Most of this book demonstrates how success in science has led to improvements in agricultural productivity, but in the case of the rabbit all that can be reported is failure. Rabbits were kept for food and fur from the time of their importation by the Normans and at a later date valued for the felt made from their hair. They were held in check because they were a valuable asset either for the use of the estate or for the letting of the warrens to others, who would harvest as many as possible. Economic and social changes in the countryside led to the multiplication of rabbits to the level of pests. Government-sponsored rabbit clubs were set up to control the rabbits by shooting (the cartridges being supplied free). The accidental introduction of myxoma virus from Australia led to massive reductions in the rabbit population and a decline in interest in shooting and ferreting, independently of the withdrawal of support for rabbit clubs by the government. The virus was transmitted by fleas in the burrows. Affected rabbits suffering from myxomatosis were badly disfigured, frequently blind and pitiable to see. Before total control became possible some rabbits began to develop behaviour patterns (building surface sets, etc.) which reduced the likelihood of infection. At the same time the virus appeared to become less virulent. The net result has been that the virus is no longer effectively controlling the rabbit population. Because of the horrific nature of the disease, attempts to introduce new virulent strains or to select new virulent strains in Britain have been discouraged. Today rabbits are major pests in large areas of the country and although they are the subject of scientific study no new, effective methods of control have emerged.

Plant diseases

Catastrophic plant disease

Plant diseases, as a rule, gain general notice only when they become catastrophic. Thus the peach yellows virus caused a catastrophic decline in peach growing in the eastern United States throughout the nineteenth century (Kunkel, 1933; Hartzell, 1935). In Ghana the cacao crop was an important currency earner and the main source of cash for the farmers of the forest belt throughout most of the twentieth century. The swollen shoot disease of cacao, a virus disease, appeared about 1936 and thereafter steadily eroded the acreage of bearing trees (Posnette, 1947). The white pine blister rust (*Cronartium ribicola*) is a famous example of a fungal disease of the white (or Weymouth) pine (*Pinus strobus*) of the eastern and central United States, where it was the main conifer forest tree. The introduction of the rust from eastern Asia and Europe and its establishment on wild *Ribes* spp. (the alternative host) in the U.S.A. had very severe consequences (Peterson and Jewell, 1968). The Dutch elm disease caused by the fungus *Ceratocystis ulmi* has caused havoc in Europe and in the United States. It was transported across the Atlantic from west to east in logs and spread by the bark beetles *Scolytus* spp. (Gibbs, 1978). *Helminthosporium oryzae*, a fungal leaf disease of rice, is generally credited with being the main cause of the devastating Bengal famine of 1943 (Padwick, 1950). The fungal rust diseases of cereals are well known and their importance as causes of devastating plant disease is appreciated. In Britain and Ireland the great famine of 1845 and 1846 caused by the failure of the potato crop, following the attack of the fungus *Phytophthora infestans*, is well documented. Other dramatic events have been the epiphytotics of fungal diseases caused by the coffee rust, *Hemileia vastatrix*, commencing in Ceylon and spreading widely; *Puccinia polysora* on maize in Africa; and Panama disease, *Fusarium oxysporum* f. sp. *Cubense*, which decimated bananas in the West Indies and Central America (Holliday, 1980).

As well as fungi and viruses, bacteria can cause serious plant diseases. Examples are fireblight of apples, pears and other rosaceous plants, caused by *Erwinia amylovora*; banana Moko wilt and ring rot of potato caused by *Pseudomonas solanacearum*; and the watermark disease of cricket bat willow, caused by *Erwinia salicis*. These are only a few examples of diseases that have been recorded as catastrophic; individual countries can demonstrate many others.

In between the epiphytotic occurrences in earlier centuries and indeed into the early years of the twentieth century, plant diseases exacted a steady toll on farmers' crops, a toll that became a major impost when the climatic circumstances dictated. On a year-by-year basis, however, British farmers did not, in those days, put plant diseases high on the list of their farming troubles. If they did, there were not many remedies at hand; they could only maintain a strict rotation of crops and grow hardy, well-proven and locally adapted varieties. The exception was the potato and potato late blight, which brought itself forcibly to farmers' attention in most years. The potato had been introduced into Europe in the sixteenth century but only became established as a farm crop in the eighteenth. Then its capacity to yield well in the wetter areas and the size of the yields allowed an increase in population and a fragmentation of croft holdings in western Scotland, and particularly in western Ireland, which made the population vulnerable to any catastrophe to the crop. That catastrophe came in the 1840s, when there was a sudden first appearance of the late blight disease (*Phytophthora infestans*). It appeared in Europe, probably from Mexico, and apparently spread from France and the Low Countries to the whole continent including Great Britain and Ireland. There was a severe outbreak in 1845 followed by an attack in 1846, which totally destroyed the crop, caused a horrendous famine and, together with outbreaks in the years immediately following, had economic and political consequences which still trouble us today (Woodham-Smith, 1962). By the twentieth century, for a variety of reasons, the disease took an annual toll but reached severe proportions in some years only. Plant diseases tend to appear in more dramatic fashion when the plant or the pathogen are transported to new countries, perhaps because transfer of either leads to new and uninhibited attack, perhaps because of the large number of small differences in the environment presented to the plant or the pathogen. In the case of late blight of potatoes the outbreak seems to have been caused by a pathogen that normally attacked wild species of Mexican potatoes while the domesticated crop plant was derived from wild species from the Andean regions of South America, which had never before been exposed to attack by this pathogen. Thus the crop plant and the pathogen met for the first time in an environment alien to both.

Prophylactic spraying

As we shall see, in modern times, the fungal diseases were the first to be controlled. To date, the application of prophylactic sprays has been much more successful with these diseases than with those caused by bacteria and viruses.

Early records of the use of copper derivatives by the Romans are hard to evaluate; the first authenticated records of the use of copper salts appear in the nineteenth century when Paris Green (copper aceto-arsenite) is recorded as a control for flea beetle and Colorado beetle. When Millardet in 1882 observed that vines treated with verdigris (copper acetate) to render them obnoxious to pilfering remained free from the vine downy mildew (*Plasmopara viticola*), a relative of late blight of potatoes, it was a simple step to try out other copper salts and to find that a mixture of copper sulphate and lime (later named Bordeaux mixture) produced a gelatinous substance that was highly protective against the vine mildew and also against potato blight. The initial difficulty was in delivering the materials evenly to the leaf surface with the hand syringes of the day. Progress to knapsack sprayers and eventually to a horse-drawn machine with several nozzles, worked by a hand pump, led to improvements in the spraying technique. But real advance depended on the development of power take-off from the tractor and the development of multiple spray heads at various levels and in various positions, depending on the crop. The problem of the quantity of water required to spray a field remained; about 150 gallons per acre (*ca.* $1750 \, \text{l ha}^{-1}$) was the usual quantity, and this had to be repeated frequently to keep the new growth of the crop covered. Tribute should be paid to J.M. Hirst and his associates, who helped spray technology by bringing precision to our knowledge of the potato blight life cycle and created an understanding of the physical factors controlling infection and epiphytotic spread (Hirst, 1955). In addition to the mechanical problems of covering the plant with an effective spray, the process of making the Bordeaux mixture was tedious and there was always a danger that the chemical reaction that removed the copper sulphate was not complete.

The cognate formulation for Burgundy mixture was made with copper sulphate and sodium carbonate. This was more likely to scorch foliage, and it resembled Bordeaux mixture in reducing the vigour of the crop when used in areas open to atmospheric pollution and also in hot dry conditions. A similar 'home-made' product was lime–sulphur, manufactured by boiling together lime and sulphur, used on fruit trees against powdery mildew. Both Bordeaux mixture and lime–sulphur could be mixed with lead arsenate or nicotine to control insect pests. All these fungicides were difficult to make up, required high-volume spraying and were likely to be toxic to the crop in particular circumstances.

Something better was required. Successful efforts were made to use copper oxychloride dusts and sprays and low-volume formulations from tractor-driven blowers or sprays or from aeroplanes, at a rate of 1–3 kg copper per hectare; for a time these seemed to be the replacements for Bordeaux mixture. But farmers were slow to replace Bordeaux mixture; as Horsfall (1975) said, it was a dead horse that would not lie down. It finally keeled over with the arrival of the first synthetic organic fungicides, the derivatives of bis-dithiocarbamic acid. Investigations into these compounds were begun by the du Pont company in 1931 and a patent obtained for their use as insecticides and fungicides in 1934. Thiram (tetramethylthiuram disulphide) was the first to be used and found its main application as a seed protectant. Disodiumethylene-bis-dithiocarbamate (Nabam) was synthesised by Rohm and Haas and reported on in 1943 (Peacock, 1978; Dimond, Heuberger and Horsfall, 1943). The zinc salt of dithiocarbamic acid, marketed as Zineb, and the manganese salt, marketed as Manzeb, were followed by Mancozeb, the double salt of manganese and zinc. With these fungicides very effective regimes were developed for the control of potato blight and other plant diseases; they remain important elements in disease control programmes to this day. As in the case of the insecticides, once the possibility of finding organic fungicides was accepted by the chemical companies, a large-scale testing programme was undertaken and new fungicides appeared regularly thereafter.

The synthetic fungicides

A great variety of new fungicides was marketed. The early compounds matched the dithiocarbamates in that they were active on the surface of the leaf. Among them were the pthalimides (polyhalogenalkylthio compounds), including such well known names as Captan and Captafol, effective against a wide range of fungi. These have proved to be exceedingly useful, although recently doubts have been raised about the health risks from some of them.

The triphenyltin compounds fentin acetate (Brestan, Hoechst) and fentin hydroxide (Du-Ter) were particularly useful against potato blight and other phycomycetes. Organo-tin compounds in general have good biocidal qualities, but only the triphenyltins combine good fungicidal activity with acceptable tolerance by the crop. They have the added advantage of a translaminar effect on potato blight. The fungicide controls the sporulating fungus on the lower surface of the leaf (where it is most plentiful) as well as on the upper surface.

All these fungicides are protective and sprayed at relatively high volume.

They are thus improvements on Bordeaux mixture only in terms of their lower phytotoxicity, a marginal improvement in fungus kill, ease of handling and, in the case of the tin compounds, their translaminar effect. They have one disadvantage in common with Bordeaux mixture; the crop grows and expands new leaves and stems, which become vulnerable to attack. Applications of fungicide must be made often enough to maintain an adequate cover.

The next breakthrough in the 1970s and 1980s was the development of systemic fungicides such as benomyl (methyl-1(butylcarbamoyl)-2-benzimidazolecarbamate), a broad-spectrum fungicide, by du Pont, cymoxanil (cyanoacetamide-oxime), and a group of phenylamides, metalaxyl, benalaxyl and oxadyxil (see Schwinn and Margot, 1991). These last are widely used against the phycomycete pathogens; they show translaminar penetration and acropetal systemic movement (i.e. towards the above-ground apices and developing leaves). None shows much basipetal movement (towards the lower leaves and roots) but fosetyl-aluminium, a phosphite (Williams *et al.*, 1977), shows both acropetal and basipetal movement and shows promise for the treatment of root diseases. In parallel, fungicides were developed particularly for the control of cereal diseases; they include the triazoles and imidazoles (e.g. propiconazole, imazalil), the pyrimidine ethirimol (Peacock, 1978), morpholines, carboxamides and others. When first introduced the systemic fungicides seemed like wonder drugs, but experience showed that the fungi could quickly change to become resistant to them. Benomyl, when used against *Botrytis cinerea* in Swiss vineyards, failed to control the disease after two years and the fungal resistance persisted thereafter. It is true to say that plant pathologists were taken by surprise by the speed of development of resistance to benomyl, but they were more prepared for its development with other systemic fungicides. They monitored for resistance in advance and combined the use of standard copper oxychloride and dithiocarbamate fungicides with, for example, metalaxyl to attempt to reduce the buildup of resistance to the valuable systemic fungicides.

The azole fungicides are active against a wide range of fungi, including powdery mildews and rusts. Wider experience has demonstrated their utility against an increasing range of fungi, including fungi attacking human beings. It has shown that their fungicidal quality lies in the demethylation inhibition of sterol biosynthesis. The main disadvantage of these compounds lies in the capacity of the fungus to develop resistance. This has been particularly marked with barley mildew. Fortunately another group of fungicides, the morpholines, which are also inhibitors of sterol biosynthesis at a different biochemical site, provide an alternative. They, however, are largely limited in the range of fungi they attack (the mildews). The older fungicide ethirimol is another alternative. With this range of fungicides, separately or in mixture, control of barley mildew seems assured. Other pathogens of cereals have shown less lability (Baldwin and Rathmell, 1988). Staub (1991) takes an optimistic view of the future use of these fungicides and believes that with appropriate care they can continue to be used for the control of the major plant diseases into the future. These fungicides stop the fungus by interfering with biochemical pathways. Other fungicides in use now or expected in the future are not directly toxic to the fungus but interfere with it by upsetting the penetration process or the appressorial attachment process.

The commercial response to the proliferation of organic fungicides has been the production of a great number of special formulations of one or more fungicides, sold under a wide variety of trade names (Schwinn and Margot, 1991). These are not only effective fungicides but are beautifully packaged, sometimes in water-soluble packaging, so that they need only be dropped into the spray tank and agitated without the operative being exposed to the dry chemical.

The improvement in spray technology, the reduction in the volume of water carrier required, and the increased speed of tractors while spraying allowed by a strengthened boom design, have allowed the extension of prophylactic spraying to new crops, especially the cereals. The concomitant development of high-input systems of cereal growing on the continent, and later in Britain, with maximum fertiliser application strategically applied throughout the growing period, and 'tram lines' (wheel spaces in the crop to accommodate tractors when the crop has grown) led to the spraying of cereals to control mildew (*Erysiphe graminis*), yellow rust (*Puccinia striiformis*), and brown rust of wheat (*P. recondita*) and of barley (*P. hordei*), together with certain other diseases. It would not be too much to claim that the efficiency and adaptability of the aminopyrimidine (ethirimol), azole and morpholine fungicides contributed greatly to this cereal revolution.

Much work has been carried out on new methods of spray application to achieve still lower volumes; development continues on air-assisted application (especially for tree crops) and various methods of charging droplets so that they are attracted to the leaf surface (e.g. the electrodyne system used in hand-held sprayers in the tropics) (Pascoe, 1985).

Certain other diseases of cereals, loose and covered smut of oats and barley, (*Ustilago* spp.), bunt of wheat (*Tilletia caries*), and leaf stripe of oats, (*Pyrenophora avenae*) and of barley (*P. graminea*) were controlled by seed dressings. Initially organomercury compounds such as cyano(methylmercury)guanidine, methylmercury 2,3-dihydroxypropylmercaptide and methylmercury acetate were used, but some posed health risks and with a general reluctance to add to the amount of mercury in the environment other compounds took over, such as Thiram (see above), triadimenol (a triazole) and carboxin (a carboxyamide), singly or in combination.

Control without spraying

There are other diseases that do not respond to prophylactic spraying but which are brought under control, indeed often disappear, as a result of the accumulated work of dedicated plant pathologists. The new knowledge is incorporated into the husbandry of the crop. One such is take-all disease of cereals, caused by *Gaeumannomyces graminis* (Garrett, 1956); it is carried between annual crops by the hibernating fungus in the dead roots and stems of the host. Garrett was able to demonstrate the cultivation factors that controlled the length of life and the infectivity of the fungus. For example, nitrogen added to the winter seed-bed prolonged the life of the pathogen, whereas nitrogen added in the spring conferred advantage on the host plant. Methods such as green manuring, which encouraged increased competition from other saprophytic fungi and bacteria, reduced the vigour and the population of the take-all fungus. With the development of husbandry systems in which wheat or barley were sown sequentially for several years in the same field it was thought that the take-all disease would increase. In fact work by Ogilvie and Thorpe (1963) and Slope *et al.* (1970) have shown that, although there will almost certainly be an increase in infection in the second year of the sequence, this is followed in later years by a decline and the disease stabilises at a reasonable level.

One other disease that has been controlled by intensive biological research work is the root and butt rot of conifers, particularly scots pine, caused by *Heterobasidion annosum*. Rishbeth's (1951) demonstration of the spread of infection in plantations by spore infection of cut stumps enabled avoiding action to be taken by protecting stumps from infection or, best of all, by artificially inoculating them with antagonistic saprophytic fungi such as *Peniophora gigantea* (Rishbeth, 1963). This simple technique has revolutionised the control of the disease in European conifer forests.

Before the second agricultural revolution, plant disease, like insect attack, was either chronic, appearing year by year to destroy a part of the crop, or sudden and cataclysmic in response to a particularly favourable weather pattern. With the development of understanding of disease cycles and the availability of powerful fungicides, some fungal diseases have tended to disappear, their control being effected by small changes in or alteration to the husbandry programme.

However, plant disease control is never static. New and hitherto unknown diseases appear. Old diseases appear in new forms. The fact that some varieties of crop plants were not attacked by the commonly occurring forms of diseases was realised long before the beginning of our revolutionary period. The fact that this 'resistance' could be inherited according to the normal laws of genetics and could be combined with other favourable agronomic qualities was probably first demonstrated by R.H. Biffen in 1907; using these principles, he later bred Little Joss wheat, resistant to the yellow rust disease caused by *Puccinia striiformis*. Sometimes a series of host varieties is attacked by a series of pathogenic forms, called races, identified in terms of the reaction of a series of varieties of the host. Vanderplank (1978) has identified this as vertical resistance (sometimes called race-specific resistance) and suggested that it is controlled by single genes and easily overcome by changes in the pathogen. On the other hand, where the relationship between the host and pathogen is less precise, where the resistance although perhaps only partial is stable and long-lasting, he has used the term horizontal resistance (sometimes called durable resistance). Many successes and also spectacular failures have resulted from the use of disease resistance in breeding programmes. Some of these are discussed further in Chapter 7.

Because of the positive interaction between crop disease and the weather, successful attempts have been made to develop forecasting systems for the onset of disease epiphytotics. A great deal of work has been done on late blight of potatoes; sophisticated systems are available which depend on such measurements as temperature, rainfall, humidity and dew-point (Bourke, 1970; Fry and Doster, 1991). Some of these attempt to forecast the onset of the epiphytotic; others, equally valuable, offer negative forecasts. An account of the recent highly sophisticated BLITECAST system is given by Fry and Doster (1991). Other predictive systems have been developed for apple scab (*Venturia inaequalis*), for rice blast (*Pyricularia oryzae*) and for many other important diseases. The savings that such programmes afford in seasons when spraying is not required, or where the spray programme can be delayed or reduced, may be considerable.

So far we have been concerned largely with fungal diseases. The symptoms of bacterial and viral diseases of plants have also been recognised for many years; the proof of association with bacterium or virus came mostly before our period. Bacterial diseases of plants were first recognised by Wakker, writing about hyacinth yellow disease in 1889, and by L.R. Jones, describing the soft rots of carrots and other plants. However, a widening appreciation of the role of bacterial diseases came in our time (Dowson, 1949) and successful control methods have been developed for important scourges of crops, particularly tropical ones. Some bacterial diseases such as angular leaf spot of cucumbers and melons (*Pseudomonas lacrymans*), which can be devastating under warm humid conditions, can be controlled by dithiocarbamate sprays. Others are controlled by strict hygiene and quarantine measures (bacterial ring rot of potatoes, *Corynebacterium sepedonicum*), by seed treatment, and by the development of resistant varieties (black arm disease of cotton, *Xanthomonas malvacearum*).

The importance of hygiene, quarantine, and regulatory measures becomes even greater with the virus diseases of crops, which have no means of dispersal from within the plant other than dispersal by (mostly sucking) insects, by dispersal in graft and cutting material (fruit trees and bushes) and occasionally by contact between plants or by human vectors carrying sap between plants (tobacco mosaic virus and some potato viruses).

The proof of association between viruses and recognisable disease symptoms was first made by Ivanowski in 1892. He showed that the sap of tobacco mosaic plants, passed through a Pasteur–Chamberland porcelain filter (known to retain bacteria), would infect healthy plants of tobacco when applied to the leaves. Thereafter the importance of virus diseases of plants became apparent. In the potato crop, for example, the practice of importing fresh stock into southern England from higher and cooler parts of the United Kingdom to replace the degenerate material was established empirically. When it was realised that virus diseases were associated with the degeneration, the (then) Department of Agriculture for Scotland instituted a scheme for the production of seed potatoes with minimal virus infection and guaranteed true to name. This was at first achieved by rogueing out the worst diseased material and any recognisable rogue varieties and then, when high-grade stocks had been obtained, multiplying from them with an annual inspection for health and purity. Later still, the possibility having emerged of freeing potatoes from viruses by heat treatment and it now being necessary to provide material free from tuber-borne fungal diseases, a system of producing virus-tested stem-cutting (VTSC) material was instituted. All potato stocks were derived from such material, new material being introduced and multiplied up at frequent intervals. The development of methods of tissue culture and micropropagation allowed the whole system to be streamlined. Schemes for soft fruit and bulbs followed; in England, the former East Malling Research Institute developed methods for indexing mother trees for scion production and for the production of virus-free root-stock material. Of course such schemes could not be followed for annual plants except where, as in the case of lettuce mosaic virus, the virus is seed-borne. In all virus diseases, which are also insect-borne, the availability of systemic insecticides helps greatly in the control of the worst spread of viruses.

Fundamental considerations

In the period of the modern agricultural revolution, although satisfactory control of many diseases was achieved by the methods outlined above, much work on diseases of plants was focused on fundamental problems. Questions concerning the process of infection, its mechanics, physiology and biochemistry, were investigated by workers such as William Brown and Ronald Wood at Imperial College, London (Wood, 1967). Others were concerned with the physiological specialisation of pathogens that existed as races attacking species or particular groups of varieties (Manners, 1950; Black, 1952; Stakman and Christensen, 1953). This work was extended by Craigie (1927), Keitt *et al.* (1948) and Shaw (1991), who examined the genetics of pathogenicity. Important work was initiated by Flor (1971) in demonstrating the gene-for-gene relationship between pathogens and their hosts. Others were intrigued by the variation exhibited by fungi that had no apparent sexual mechanism to explain their capacity for frequent variation.

Pontecorvo (1956) explained this by the parasexual cycle where nuclear combination, followed by meiosis and segregation, apparently took place in the vegetative phase of the *Penicillia*. Buxton (1956) extended this concept to the pathogenic fungus *Fusarium oxysporum*.

The study of pathogens that could not survive apart from their living host intrigued workers for many years and was finally consigned to history when Scott and Maclean (1969) demonstrated the growth of cereal rust fungi in artificial culture. Similarly, the problem of inducing fruiting of *Phytophthora* spp. in artificial culture was finally solved when no less than four groups of workers reported independently within a period of six months that sterols were necessary for the production of oospores of these fungi. Equally exciting work was being carried out on the plant viruses. These, at first mysterious, entities were shown by Bawden and Pirie at Rothamsted (1937) and Stanley at Princeton (1936) to be extractable. The extract could be purified by chemical means and then inoculated to the host, where it would multiply and cause symptoms; from these reinfected plants the virus could be extracted once more. This was the opening to further investigations into the nature of the nucleic acid constituents of the viruses and in a sense the very foundation of the modern science of molecular biology.

These findings (and many others not reported here) formed a whole infrastructure, which currently supports the present-day research work, which seeks to provide sophisticated means of controlling pathogen attack on crop plants. The central problem is the elucidation of the mechanism of recognition of the host by the pathogen and of the pathogen by the host, and the consequent biochemical reactions leading to exclusion of the pathogen on incompatible hosts (Müller and Behr, 1949; Ingram, 1967; Friend, 1991). A deeper understanding of this phenomenon must be the key to new methods of plant protection. It might be achieved by providing biochemical analogues to trigger the initiation of exclusion mechanisms. Alternatively, with deeper understanding it should be possible to breed crop plants that maintain a high level of resistance over a prolonged period.

As in the case of weed and insect control it is possible to put a monetary value on the savings that modern methods of control provide (Schwinn and Margot, 1991; Zadoks, 1985). Such exercises are useful but soon become dated. In real-life farming, local price fluctuations, differing values between individual crops (wheat in demand, barley not) and between crops and stock (up corn, down horn) and the spread of costs across enterprises makes evaluation difficult. In the case of late blight of potatoes, wet years are blight years. This in turn means high cost of spraying, damaged tubers and more discards at the riddle, but high yields (because of the wet growing conditions) and, even with the reduction in yield brought about by blight, a crop that is too large for the market and which without the intervention of the Potato Marketing Board (which regulates the market by buying excess production for stock feed and is currently under threat) would suffer severe reductions in price.

The value of weed, pest and disease control is as much in the increased knowledge and security it gives the farmer, as in the value of the increased and more saleable crop. This security translates into a control of the annual production which was absent in the days before the second agricultural revolution. Farmers can plan their businesses with greater certainty, confident that, apart from exceptional natural disasters, they can budget for the production of expected yields at economic prices and attempt to run a business, not a lottery on the fickleness of the weather.

Selected references and further reading

Baldwin, B.C. and Rathmell, W.G. (1988). Evolution of concepts for chemical control of plant disease. *Annual Review of Phytopathology* **26**, 265–83.

Bawden, F.C. and Pirie, N.W. (1937). The isolation and some properties of liquid crystalline substances from solanaceous plants infected with three strains of tobacco mosaic virus. *Proceedings of the Royal Society, London* B **123**, 274–320.

Biffen, R.H. (1907). Studies in the inheritance of disease resistance. *Journal of Agricultural Science* **2**, 109–28.

Black, W. (1952). Inheritance of resistance to blight (*Phytophthora infestans*) in Potato: inter-relationships of genes and strains. *Proceedings of the Royal Society of Edinburgh* B **64**, 312–52.

Bourke, P.M.A. (1970). Use of weather information in the prediction of plant disease. *Annual Review of Phytopathology* **8**, 345–70.

Buxton, E.W. (1956). Heterokaryosis and parasexual recombination in a pathogenic strain of *Fusarium oxysporum*. *Journal of General Microbiology* **15**, 133–9.

Carson, Rachel (1963). *Silent Spring*. Hamish Hamilton, London.

Charnley, A.K. (1989). Mycoinsecticides, present use and future prospects. In *Progress and Prospects in Insect Control, Proceedings of an International Conference held at the University of Reading*, ed. N.R. Macfarlane, pp. 131–44. British Crop Protection Council, Farnham, England.

Conway, G.R. & Pretty, J.N. (1991). *Unwelcome Harvest: Agriculture and Pollution*. Earthscan Publications, London.

Craigie, J.H. (1927). Discovery of the function of the pycnia of the rust fungi. *Nature, London* **120**, 765–7.

Dimond, A.E., Heuberger, J.W. & Horsfall, J.G. (1943). A water soluble protective fungicide with tenacity. *Phytopathology* **33**, 1095–7.

Darwin, C. & Darwin, F. (1880). *The Power of Movement in Plants*. John Murray, London.

Dodd, A.P. (1940). *The Biological Campaign against Prickly Pear*. Commonwealth Prickly Pear Board, Brisbane, Australia.

Dowson, W.J. (1949). *Manual of Bacterial Plant Diseases*. Adam and Charles Black, London.

Elliot, M. (1989). Pyrethroids, past, present and future. in *Progress and Prospects in*

Insect Control, Proceedings of an International Conference held at the University of Reading, ed. N.R. Macfarlane, pp. 43–4. British Crop Protection Council, Farnham, England.

Entwistle, P.F., Adens, P.H.W., Evans, H.F. and Rivers, C.F. (1983). Epizootiology of a nuclear polyhedrous virus (Baculovirideae) in European Spruce Sawfly (*Gilbinia hercyniae*). *Journal of Applied Ecology* **20**, 473–87.

Flor, H.H. (1971). Current status of the gene-for-gene concept. *Annual Review of Phytopathology* **9**, 275–96.

Friend, J. (1991). The biochemistry and cell biology of interaction. In *Advances in Plant Pathology*, vol. 7, ed. D.S. Ingram and P.H. Williams, pp. 85–129. Academic Press, London.

Fry, W.E. and Doster, M.A. (1991). Potato late blight: forecasts and disease suppression. In *Phytophthora; a Symposium*, ed. J.A. Lucas, R.C. Shattock, D.S. Shaw and Louise R. Cooke, pp. 326–36. Published for the British Mycological Society by the Cambridge University Press.

Fryer, J.D. and Evans, S.A. (1968). *Weed Control Handbook of the British Crop Protection Council* (two volumes), 5th Edition. Blackwell Scientific Publications, Oxford.

Fryer, J.D. and Makepeace, R.J. (1978). *Weed Control Handbook of the Crop Protection Council, Recommendations*. 8th Edition, vol. 2. Blackwell Scientific Publications, Oxford.

Garrett, S.D. (1956). *Biology of Root-infecting Fungi*. Cambridge University Press.

Gibbs, J.N. (1978). Intercontinental epidemiology of Dutch Elm Disease. *Annual Review of Phytopathology* **16**, 287–307.

Grossbard, E. and Atkinson, D. (eds) (1985). *The Herbicide Glyphosate*. Butterworths, London.

Hance, R.J. and Holly, K. (1990). *Weed Control Handbook of the British Crop Protection Council, Principles*. 8th Edition, vol. 1. Blackwell Scientific Publications, Oxford.

Hannay, C.L. (1953). Crystalline inclusions in aerobic spore-forming bacteria. *Nature, London* **172**, 1004.

Harper, F. (1983). *Principles of Arable Crop Production*. Granada Publishing, London.

Hartzell, A. (1935). A study of Peach Yellows Virus. *Contributions from the Boyce Thompson Institute* **7**, 183–207.

Hewson, R.T. and Roberts, H.A. (1971). The effect of weed removal at different times on the yield of bulb onions. *Journal of Horticultural Science* **46**, 471–83.

Hirst, J.M. (1953). Changes in atmospheric spore content: diurnal periodicity and the effects of weather. *Transactions of the British Mycological Society* **36**, 375–92.

Hirst, J.M. (1955). The early history of a potato blight epidemic. *Plant Pathology* **4**, 44–50.

Holliday, P. (1980). *Fungus Diseases of Tropical Crops*. Cambridge University Press.

Horsfall, J.G. (1975). The story of a non-conformist. *Annual Review of Phytopathology* **13**, 1–13.

Hussey, N.W. (1990). Agricultural production in the third world. A challenge for natural pest control. *Experimental Agriculture* **26**, 171–83.

Ingram, D.S. (1967). The expression of R-gene resistance to *Phytophthora infestans* in tissue cultures of *Solanum tuberosum*. *Journal of General Microbiology* **49**, 99–108.

Jones, D.G. and Clifford, B.C. (1978). *Cereal Diseases*. BASF U.K. Ltd., Ipswich.

Jutsum, A., Poole, N.J., Powell, K.A. and Bernier, R.L. (1989). Insect control using microbial toxins; status and prospects. In *Progress and Prospects in Insect Control*,

Proceedings of an International Conference held at the University of Reading, ed. N.R. Macfarlane, pp. 131–44. British Crop Protection Council, Farnham, England.

Keitt, G.W., Leben, C. and Shay, J.R. (1948). *Venturia inaequalis*, IV. Further studies on the inheritance of pathogenicity. *American Journal of Botany* **35**, 334–6.

Kirkwood, R.C. (1991). *Target Sites for Herbicide Action.* Plenum Press, New York and London.

Kögl, F. & Haagen Smit, A.J. (1931). Uber die Chemie des Wuchstoffs. *Proceedings Koninklijke Nederlands Akademie van Wetenschappen te Amsterdam* **34**, 1411–16.

Kunkel, L.O. (1933). Insect transmission of Peach Yellows. *Contributions from the Boyce Thompson Institute* **5**, 19–28.

Leakey, J.P. (ed.) (1985). *The Pyrethroid Insecticides.* Taylor and Francis, London.

Manners, J.G. (1950). Studies on the physiologic specialization of Yellow Rust (*Puccinia glumarum*) in Britain. *Annals of Applied Biology* **37**, 187–214.

Meehan, A.P. (1984). *Rats and Mice: their Biology and Control.* Rentokil Ltd., East Grinstead, U.K.

Mellanby, K. (1989). D.D.T. in perspective. In *Progress and Prospects in Insect Control, Proceedings of an International Conference held at the University of Reading*, ed. N.R. Macfarlane, pp. 3–20. British Crop Protection Council, Farnham, England.

Mellanby, K. (1992). *The DDT Story.* British Crop Protection Council, Farnham, England.

Müller, K. and Behr, L. (1949). Mechanism of Phytophthora-resistance of potato. *Nature, London* **163**, 498–9.

Naumann, K. (1989). Acetylcholinesterase inhibitors. In *Progress and Prospects in Insect Control, Proceedings of an International Conference held at the University of Reading*, ed. N.R. Macfarlane, pp. 21–41. British Crop Protection Council, Farnham, England.

Nutman, P.S. (1963). Factors influencing the balance of mutual advantage in legume symbiosis. In *Symbiotic Associations, the Thirteenth Symposium of the Society for Microbiology*, ed. P.S. Nutman and Barbara Mosse, pp. 51–71. Cambridge University Press.

Ogilvie, L. and Thorpe, I.G. (1963). The relation of disease control to successful continuous cereal growing. *N.A.A.S. Quarterly Review* **58**, 65–9.

Ordish, G. (1976). *The Constant Pest.* Peter Davies, London.

Padwick, G.W. (1950). *Manual of Rice Diseases.* Commonwealth Mycological Institute, Kew, Surrey, England.

Pascoe, R. (1985). Biological results obtained with the hand held 'Electrodyn' spraying system. In *Application and Biology*, ed. E.S.E. Southcombe, pp. 75–85. Monograph No. 28, British Crop Protection Council, Farnham, England.

Peacock, F.C. (1978). *Jealott's Hill – Fifty Years of Agricultural Research, 1928-78.* Imperial Chemical Industries Ltd., Bracknell, U.K.

Peterson, R.S. and Jewell, F.F. (1968). Status of American stem rusts of pine. *Annual Review of Phytopathology* **6**, 23–40.

Pontecorvo, G. (1956). The parasexual cycle in fungi. *Annual Review of Microbiology* **10**, 393–400.

Posnette, A.F. (1947). Virus diseases of cacao in West Africa. I. Cacao virus IA, IB, IC, and ID. *Annals of Applied Biology* **34**, 388–402.

Putter, I., MacConnell, J., Preiser, F.A., Hardvic, A.A., Ristrich, S.S. and Dybus, R.A. (1981). Avermectins: novel insecticides, acaricides and nematocides. *Experientia* **37**, 963–4.

Ratcliffe, D.A. (1970). Changes attributable to pesticides in egg breaking frequency and eggshell thickness in some British birds. *Journal of Applied Ecology* 7, 67–115.

Rishbeth, J. (1951). Observations on the biology of *Fomes annosus* with special references to East Anglian pine plantations. II. Spore production, stump infection and saprophytic activity in stumps. *Annals of Botany, London* 14, 365–83.

Rishbeth, J. (1963). Stump protection against *Fomes annosus*. III. Inoculation with *Peniophora gigantea. Annals of Applied Biology* 52, 63–77.

Rogers, Heather (1992). Screwworm wiped out in North Africa. *Shell Agriculture* 12, 26–8.

Rudd Jones, D. (1981). Science in horticulture. In *Agricultural Research 1931-1981*, ed. G.W. Cooke, pp. 162–182. Agricultural Research Council, London.

Russell, E.J. (1966). *A History of Agricultural Science in Great Britain 1620-1954.* George Allen and Unwin, London.

Scott, K.J. and Maclean, D.J. (1969). Culturing of Rust Fungi. *Annual Review of Phytopathology* 7, 123–46.

Sanders, H.G. (1943). *An Outline of British Crop Husbandry.* Cambridge University Press.

Schwinn, F.J. and Margot, P. (1991). Control with chemicals. In *Phytophthora infestans, the cause of late blight of potatoes* (*Advances in Plant Pathology*, vol. 7), ed. D.S. Ingram and P.H. Williams, pp. 225–65. Academic Press, London.

Shaw, D.S. (1991). Genetics. In *Advances in Plant Pathology*, vol. 7, ed. D.S. Ingram and P.H. Williams, pp. 131–70. Academic Press, London.

Slope, D.B., Etheridge, J., Henden, D.R. and Hornby, D. (1970). Take-all development and decline. *Report of Rothamsted Experiment Station* 1, 158–9.

Stakman, E.C. and Christensen, J.J. (1953). Problems of variability in fungi. In *Plant Diseases – The Yearbook of Agriculture*, ed. A. Stefferud, pp. 35–62. U.S. Department of Agriculture, Washington, D.C.

Stanley, W.M. (1936). Chemical studies on the virus of tobacco disease. VI. The isolation from diseased Turkish tobacco of a crystalline protein possessing the properties of tobacco-mosaic-virus. *Phytopathology* 26, 305–20.

Staub, T. (1991). Fungicide resistance: practical experience with anti-resistance strategies and the role of integrated use. *Annual Review of Phytopathology* 29, 421–42.

Templeman, W.G. and Sexton, W.A. (1946). The differential effect of synthetic plant growth substances and other compounds upon plant species. I and II. *Proceedings of the Royal Society, London* B 133, 300–13; 480–3.

Urech, P.A., Schwinn, F.J. and Staub, T. (1977). A novel fungicide for the control of late blight, downy mildews and related soil borne diseases. *Proceedings 9th British Crop Protection Conference on Pests and Diseases* 2, 623–31.

Vanderplank, J.E. (1978). *Genetic and Molecular Basis of Plant Pathogenesis Assessment.* Springer-Verlag, Berlin.

Went, F.W. (1928). Wuchsstoff und Wachstum. *Recueil des Travaux Botaniques Neerlandais* 25, 1–116.

WHO/FAO (1973). Expert Group GTRES. The use of viruses for the control of insect pests. *World Health Organisation Technical Report*, Series 531.

Williams, D.J., Beach, B.G.W., Horriere, D. and Marechal, G. (1977). LS 74-783, A new systemic fungicide against phycomycete diseases. *Proceedings 9th British Crop*

Protection Conference, on Pests and Diseases **2**, 565–73.

Wood, R.K.S. (1967). *Physiological Plant Pathology* (*Botanical Monographs*, vol. 6.) Blackwell Scientific Publications, Oxford and Edinburgh.

Woodham-Smith, Cecil (1962). *The Great Hunger. Ireland 1845-1849*. Hamish Hamilton, London.

Zadoks, J.C. (1985). On the conceptual basis of crop loss assessment. The threshold theory. *Annual Review of Phytopathology* **23**, 455–73.

7 Breeding more productive plants

Plant breeding, which in 1936 was a simple art depending on serendipity and the breeder's instinct, had become, by the end of our fifty-year period, a highly sophisticated application of the science of botany to the manipulation of the plant's genetic capacity. Botany itself has evolved over the same period into a number of rapidly advancing branches of pure science, plant genetics, physiology, biochemistry, pathology, molecular genetics, morphology, cytology and taxonomy; parts of each of these have become sharply focused on the practical problem of improving crop plants for human use. As the science has increased in power, so that there is scarcely any goal that seems beyond reach, countervailing forces from the inertia of the agricultural market, with its capacity to evaluate the weaknesses as well as the strengths of apparently promising varieties, have increased.

Advances in plant breeding have taken place world-wide, but in this chapter work undertaken in the United Kingdom will be discussed particularly, because the complex interactions between breeding and agricultural change can be understood best on the smaller canvas.

British agriculture in our revolutionary period has utilised a very limited number of crops. Wheat and barley were the major cereals throughout; the oat crop was important at the beginning of the period but has since declined rapidly; rye has never been of other than limited importance on light dry soils. Cash crops were represented by sugar beet and potatoes. Fodder for livestock, apart from cereals, was provided by swedes, turnips, kale or rape, and mangels. Grazing and preserved fodder were contributed by grasses and clovers and, in particular areas of the country, by lucerne and sainfoin. There has also been a move to embrace, as large-scale horticultural crops, peas and faba beans, which previously were only grown on a field scale for fodder, and the variety of brassica crops previously grown on a market-garden scale. As we see in Chapter 8, oil seed rape and fodder maize became important towards the end of the period.

A little history

The small-grained cereals, the fulcrum of the farming system, responded well to the efforts of the early plant breeders. Their pollination mechanisms ensured a high degree of self-fertilisation, so that a few (perhaps seven or eight) generations of selection led to a reasonable degree of genetic stability. In the eighteenth century they occurred as landraces, local adaptations to particular regions; there are many tales of the selection of superior ears in a crop, which after multiplication gave rise to new and higher-yielding stocks, then designated as varieties. Informed plant breeding, however, had to await the rediscovery and digestion of Mendel's work on the principles of genetics by Bateson in 1900. The first informed crossing in the U.K. was performed by Rowland Biffen, a student of Bateson, when in 1905 he described segregation in the second generation of a cross between a number of wheat varieties including, surprisingly, Rivet and Red King (surprisingly because Rivet is a tetraploid wheat (*Triticum turgidum*) while Red King is a hexaploid bread wheat (*T. aestivum*) and segregation in their crosses is likely to be irregular) and in 1907 the segregation for yellow rust resistance in crosses between the varieties American Club (very resistant) and Michigan Bronze (very susceptible). In 1910 he introduced his variety Little Joss, derived from a cross between the English variety Squarehead's Master and a yellow-rust-resistant Russian wheat named Ghirka; it had good agronomic qualities and yellow rust resistance. In 1916 he introduced the variety Yeoman, combining the agronomic qualities of the English variety Browick with the excellent breadmaking qualities of the Canadian variety Red Fife. The useful life of this variety extended well into the middle of our period (Lupton, 1987). In 1912 the University of Cambridge set up the Cambridge University Plant Breeding Institute with Biffen as its Director. This was later to become the Agricultural Research Council's Plant Breeding Institute. This little bit of history is included here to serve as a benchmark for the later developments in plant breeding in the United Kingdom.

Wheat in Britain

The early English wheats were largely autumn-sown (winter wheats), long-strawed, and in general not noted for their bread-making quality. Yellow rust (*Puccinia striiformis*) was a frequent problem, building up on the winter phase, especially during a mild autumn. In British wheats the disease produced an annual loss then estimated at 5–10%; in imported wheat varieties grown in Britain it could cause much more severe losses. The early breeders set themselves the task of producing higher-yielding wheats with increased disease resistance and better bread-making quality, hence, as already mentioned, Biffen's red-grained Little Joss and Yeoman and Engledow's white-grained Holdfast (with good bread-making quality but spoiled by a tendency to sprout in the ear, derived from its parent, White Fife). These early breeders succeeded because they used varieties field-tested against yellow rust and in spite of their ignorance of the detailed genetics of rust resistance (Manners, 1950). Similarly, they knew about bread-making quality, although ignorant of the detail of its genetics (discussed later in this chapter). The early wheats and their immediate successors were long-strawed. Wheat straw was a valuable part of the farming economy, used as bedding and as thatch for stacks. Perhaps, too, it was useful to over-top the weed crop, which at that time could be controlled only with difficulty.

To see the development of wheat in perspective we have to remember that in the seventeenth and eighteenth centuries the bread of the general populace was heavier and coarser than the bread we know today, and milling techniques were still evolving. Barley, which contains some gluten in the endosperm, was also sometimes used as a component of the flour for bread-making. Wheats of various qualities undoubtedly existed in Britain and bread-making wheats might be imported from the continent for luxury use, but an internal market in high-quality wheat was scarcely developed. It was only with the repeal of the Corn Laws (1846) and the first importations of American and Canadian grain in 1875 that it became possible to make a beginning to the standardisation of bread in a fashion which we would recognise today. The development of industrial towns with relatively affluent inhabitants formed a ready market for a standard bread loaf. The easy availability of 'hard' wheat from Canada and the U.S.A., which were high in gluten and of superb bread-making quality, meant that there was little call for the production of bread-quality wheats in Britain. Most of the wheats grown on British farms came to be soft wheats for animal feed and for cake- and biscuit-making, where rising if required is provided by, for example, eggs and baking powder. In these flours high gluten content is a disadvantage. There followed, in history, a tug-of-war between free trade and protectionism and a rearguard action by plant breeders, hopeful that the production of a suitable bread-quality wheat for British conditions would open a market for such wheat in British agriculture. In fact, as we shall see later, it was not until the institution of an import levy on bread-quality wheat (1976) under the E.E.C. Common Agricultural Policy that progress was made in the realisation of this dream. The last element in the rearguard action was when Bell and Lupton instituted a programme in 1948 at the Cambridge Plant Breeding Institute aimed at the production of high-yielding wheats of bread-making quality, such as Maris Widgeon. This programme came into its own under the E.E.C. Common Agricultural Policy.

However, from 1936, the date of the release of Engledow's Holdfast, wheat breeders were, in general, seeking to improve yield and agronomic qualities regardless of bread-making quality. Continental varieties such as Cappelle-Desprez and Professeur Marchal were extensively grown alongside a limited range of British-bred wheats. At that time the problem of lodging (falling over) of the crop was an important element in the breeders' thinking. Yield responses to increased fertiliser application could not be realised fully because of lodging; at any level of fertiliser treatment, lodging was likely to cause losses, reduced grain quality, and harvesting difficulties in most years. Since it was now less important to have long straw for bedding and thatching, breeders were free to seek to incorporate dwarfing genes from a variety of sources. We can see a steady reduction in the height and increase in the yield of varieties put out by the Plant Breeding Institute from 1967 to the end of our fifty-year period. Some of these varieties owed their reduced stature to the selection of quantitative characters in the genetic background and others to single identifiable genes for dwarfness, the most notable of which were the genes *Rht1* and *Rht2* derived from the Japanese variety Norin 10. (The *Rht3* gene from the variety Tom Thumb was never used because of its severe side effects.) It soon became apparent that the introduction of these dwarfing characteristics brought an added bonus in the form of increased yield. These new products were, however, feed wheats. The hope that a British wheat of bread-making quality might be produced persisted. In 1979 and 1980 the semi-dwarf varieties Bounty and Avalon were released, the latter being rapidly accepted as a useful bread-making quality wheat of moderately high yield. It should also be noted that the agronomic advances in the timing of nitrogen fertiliser application have been helpful to the improvement of protein content of wheat grain.

Two other factors favoured the acceptance of these bread-making quality wheats. The first was the adoption by the E.E.C. of levies on wheat imports from outwith the Community, in 1976–77, effectively placing a premium on home-grown wheats suitable for bread-making. The second was the development of the Chorleywood process of bread-making, where an input of mechanical power in the mixing process ensured very even mixing, developed the extensibility of the dough, incorporated tiny air bubbles in the dough (which later became the focus for the accumulation of carbon dioxide from the fermentation), and produced a denser and heavier 'bread-piece', allowing economies of space and time in the baking process. It had the added advantage that it allowed the use of a larger proportion of British wheats in bread-making flours (Barnes, 1989).

Bread-making quality has long been recognised but its physiological and genetical base became more apparent when P.I. Payne at the Plant Breeding Institute showed that all the varieties of wheat selected there for bread-making quality possessed a glutenin protein subunit, which had been derived from the Canadian variety White Fife (one of the parents of Holdfast) and which genetic analysis showed to be controlled by a gene on chromosome 1A. Electrophoretic separation of the protein components of the wheat grain allowed the identification of the glutenin subunit and encouraged further searching. The result has been the identification of further subunits of glutenin controlled by genes situated on this chromosome and its homologues, with extensive variation in the related alleles. The study has been extended to the gliadins and other proteins of the wheat endosperm and to the protein components of wild diploid relatives of the bread wheat, which are thought to be valuable for future breeding programmes. Since the tests can be made with a fragment of a single grain, the mechanics of the breeding programme is thereby simplified.

Yields of modern wheat varieties

These modern wheat varieties yield about 50% more grain than their long-strawed progenitors. As Riley (1981) points out, when grown under the same conditions the mass of plant material of older and newer varieties harvested per hectare has not changed (Table 11). What has changed is the harvest index (the percentage of the total above-ground portion of the crop harvested as grain). Riley's figures also demonstrate that although the variety Norman yields 50% more than the older variety Little Joss, the yield of the latter, at approximately 6 t ha^{-1}, is still high when grown under the same conditions as the modern varieties with all the agronomic inputs of nitrogen, spacing, spraying and, in this instance, support by the use of wide-mesh netting to prevent lodging. The quoted national average yields for early wheats under British farming conditions in 1948 were about 2.5 t ha^{-1}; by 1975 the national average yield was twice that tonnage. Under the best management the new varieties could achieve four times the 1948 average yield, and exceptional crops are recorded with six times that figure. Much of the credit for these advances goes to F.G.H. Lupton and J. Bingham, who provided many of the new varieties, beginning with Maris Huntsman in 1972.

Improved plant breeding and improved agronomic management have gone hand in hand, one facilitating the other, so that it is difficult to separate out their individual contributions by a superficial analysis. In an interesting analysis of crop statistics, Valerie Silvey (1978) has estimated that 60% of the increased yield, 2.04 t ha^{-1} over the period 1947–75, could be attributed to varietal improvement, the rest relating to agronomic progress.

Physiological considerations

The redistribution of the plant's resources, less stem and more ears, shown by the change in the harvest index, must be reaching its limit; a deeper understanding of the plant's physiology is required to allow further progress. Already we know that the theoretical limit of the yield of single plants grown under the best possible conditions, with optimal light, water and fertiliser regimes, far exceeds anything so far achieved (Penman, 1968). Some quite simple advances in physiological understanding could bring about improvements in the success of the breeding programme. For example, work by R.B. Austin (1987) on the physiology of plants in the field, under normal cropping conditions, demonstrated that the assimilated carbon can move freely from tiller

Table 11. *Comparison of varieties of wheat of different ages in a trial at the Plant Breeding Institute, Cambridge, in 1977–78*

Variety (year of introduction)	Grain yield (t ha^{-1})	Height (cm)	Harvest index (%)
Little Joss (1908)	6.7	142	36
Holdfast (1935)	5.9	126	36
Maris Huntsman (1972)	7.8	106	46
Norman (1980)	9.0	84	51

Source: After Riley (1981), p. 122.

to tiller in the early stages of crop growth. In the later stages, the tillers become virtually autonomous. This information has been incorporated into the criteria for breeding and selection. Varieties that show 'tiller economy', producing fewer tillers, a larger proportion of which give rise to ears, are thought to use the carbon assimilated to better agronomic advantage at appropriate seed rates. Austin has also shown that at an early stage of ear emergence soluble carbohydrates accumulate in the still elongating stem. This carbohydrate is mobilised and moves to the grain in the last half of the grain-filling period. In a drought period the temporarily stored assimilate can contribute as much as half the carbon for grain growth. Assimilate is also produced for the development of the grain from the flag leaves and from the spikelets themselves. The interactions between these activities is complex; they had not been incorporated into the plant breeder's criteria by the end of our period.

Theoretically (Fowden, 1981) in a modern 10 t ha^{-1} wheat the yield of grain and straw is equivalent in energy terms to about 1% of the total annual solar radiation, apparently a very inefficient system. If the plant breeder knew what controlled the efficiency of the system it might be possible to improve still further the yields of future varieties. Monteith (1977) has enumerated the points at which efficiency can be seen to be lost.

Firstly, the architecture of the crop canopy has been selected by people and by nature in response to a number of environmental pressures to develop characteristics which may now be irrelevant. More erect leaves in the crop canopy are theoretically more efficient. This aspect has been tackled by T.T. Chang and P.R. Jennings at the International Rice Research Institute in the Philippines. They developed an erectoid form for rice with good yields. The erectoid form of barley has been incorporated into varieties such as Jupiter (1978), derived from Midas, which proved to be very high-yielding but which did not persist for other agronomic reasons. Short erect plants allow a greater penetration of light into the canopy and may have additional benefits in reducing humidity and disease. With such advantages one wonders what prevented their selection before. One reason might be that they do not suppress weeds, such as volunteer potatoes, which can on occasion overtop them.

Secondly, Monteith points to the quantum nature of light and the ineffectiveness of certain wavelengths of light. At present there seems no way in which these factors could be improved under field conditions.

Thirdly, the light saturation of the photosynthetic system is constrained by low ambient carbon dioxide levels. This problem is overcome under glasshouse conditions by the injection of carbon dioxide into the enclosed atmosphere, and it may be that the increase in carbon dioxide in the atmosphere, postulated to result from increased human activity, will favour plant growth in the future.

Fourthly, photosynthetically fixed carbon is lost in dark respiration.

Finally, there is a loss of assimilate in photorespiration.

These last two constraints offer no plant breeding possibilities at the present stage of technology but once identified offer a challenge to the ingenious breeder.

A pre-requisite for the development of a breeding programme, particularly with a complex hexaploid like bread wheat (with 42 chromosomes), was an understanding of the cytological basis of the breeding system. Knowledge of chromosome morphology allowed the experimental synthesis of amphidiploids from diploid and tetraploid relatives of the bread wheat by crossing with wild genera. These in turn could be crossed with the bread wheat, thus introducing qualities such as disease resistance, protein quality, etc., into its genome, but frequently also introducing undesirable characters. This was followed by Riley's work on chromosome substitution and deletion, which was greatly helped by his discovery, along with V. Chapman, of the role of chromosome 5B in controlling meiotic pairing. A haploid wheat plant lacking the 5B chromosome showed remarkably high rates of chromosome pairing at meiosis. Further work showed that the presence of this chromosome or of the gene *Ph*, which it carried, prevented pairing of homologous (equivalent) chromosomes. The removal of the gene allowed the pairing of chromosomes from genera such as *Agropyron* and *Aegilops* with equivalent chromosomes from wheat. The full exploitation of this discovery depended on a simple but effective methodology which was developed for transferring chromosomes and part chromosomes into the wheat genome. Accidental loss of a chromosome is not uncommon in plants, where it is called aneuploidy. By assiduous screening, a series of aneuploids was built up containing examples of the loss (monosomic), or addition (trisomic and tetrasomic) of each of the constituent chromosomes and identifiable part chromosomes. Such a series was developed by Sears in America for the variety Chinese Spring, and at Cambridge by C.N. Law for Holdfast, Maris Hobbit and Cappell-Desprez. Eventually the search for aneuploids was pursued on a European basis in many varieties with the formation of the European Wheat Aneuploid Cooperative.

Using aneuploids, it was possible to subtract and add appropriate chromosomes and particular genetic qualities in a plant breeding programme, thus facilitating the transfer of genetic material for specific purposes between varieties and between species. Although use is made of these techniques at various times within a plant breeding programme, the full exploitation of these aspects of the plant's physiology and cytology will be in the future.

Another discovery that awaits exploitation is the recently identified difference in the assimilation of carbon dioxide in two distinct categories of plants. In the C_3 category, which includes most temperate crop plants including wheat, barley and oats, the first step in the assimilation of carbon dioxide is the addition of carbon dioxide to a five-carbon sugar to give two molecules of phosphoglyceric acid (three-carbon), which are reduced to give six-carbon sugars. However, at the same time as carbon dioxide is being assimilated, a parallel activity using the same enzyme (ribose diphosphate carboxylase) adds oxygen to the five-carbon sugar with the formation of one molecule of phosphoglyceric acid and one molecule of phosphoglycollic acid (two-carbon), which is oxidised with the release of carbon dioxide (photorespiration). The C_4 category occurs in certain highly productive tropical crops such as maize and sorghum. It is characterised by a first step where carbon dioxide is added to phosphoenolpyruvic acid to give a four-carbon malate anion, which forms the basis for further elaboration of carbohydrate storage material. The important points arising from this necessarily rather abbreviated exposition of the two processes are: (a) that the C_3 plants appear to be made less efficient by virtue of their photorespiration and that these plants become photosaturated at about 25% of full sunlight, at normal carbon dioxide concentrations; and (b) that the C_4 plants do not suffer from the inefficiency of photorespiration and that photosynthesis continues to increase with increasing light intensity, to near-full sunlight (Chollet and Ogden, 1975).

It is clear that at present we have little ability to exploit these differences; even with the advances in molecular genetics we can already envisage, it seems unlikely that progress in this direction will be easy. It is also worth considering why, if the C_4 pathway is so much more efficient in warm climates, it has not become the major photosynthetic system for plants in cooler climates.

We have seen something of the shifting agricultural background and the advancing techniques of pure science against which the wheat breeding programme was designed. The importance of an increasing knowledge of the fundamental sciences for present and future programmes is obvious; new knowledge is crowding in all the time. The techniques of molecular biology, for example, offer possibilities of manipulation of the genome far beyond anything we can contemplate at the moment. Such genetically engineered plants, however, will still have to be integrated into farming systems and markets that are very demanding. One might envisage that the production of varieties by these modern methods might be slow at first, for technical and regulatory reasons, and that further study of the gross physiology of plants, such as the development of salt tolerance, will provide new practical challenges for the plant breeder using, at least in part, the old classical methods.

Barley breeding

The barley breeding programme differed from that of wheat in a number of respects. Barley (*Hordeum vulgare*) and wheat are known to have originated in the Fertile Crescent of the Near East and spread into north-west Europe, with barley appearing on the shallower soils and in climatically less favoured areas. Curiously, it is considered that both brought with them weed species, rye accompanying wheat and oats accompanying barley; these, in turn, moved out onto even less favoured sites (J.G. Hawkes, 1983). Barley was used as a bread grain as well as for animal feed in areas where it was the primary cereal, but it was and is used extensively for brewing and distilling throughout its range. It is a diploid plant and therefore is easier to handle in a breeding programme than wheat, but like wheat it is largely self-fertile and genetically stable. Cultivated barley is thought to be the domesticated descendant of *Hordeum spontaneum*, which occurs commonly in a variety of forms in well-marked ecological plant groupings throughout the Near East. Traditionally the cultivated barleys have existed as two-rowed (the most widely grown), four-rowed and six-rowed varieties, depending on the floret arrangement in the ear. The four-rowed forms were called bere; until recently, they were extensively grown in northern Scotland and in the climatically less favoured areas of north-west Europe. The six-rowed form, which is thought to have originated early in the evolution of the crop, includes winter hardy forms which were occasionally seen as an autumn-sown catch crop, being cut green for silage in the early summer, or folded with stock. It is now a source of winter hardiness in barley breeding programmes. Neither four- nor six-rowed barleys were useful for malting commercially. When oats were grown for stock-feed, and as a major component of the feed for horses, barley was used to a lesser extent in animal rations. Oats were probably less demanding, certainly in terms of soil acidity and in adaptability to a range of climatic conditions, and were higher-yielding under the conditions prevailing on most arable farms in northern and western Britain. However, the decline in horse numbers and the appearance of Danish stiff-strawed feeding barleys changed the situation, and barley joined wheat as an important feed grain on British farms. Throughout, barley had been grown extensively for malting, for which a premium could be obtained; one of the great skills of an arable farmer in the appropriate area was the regular production of high-quality crops for malting.

Winter and spring malting barleys

The number of varieties available between the First and Second World Wars was small; they were moderate yielders, rather long-strawed, and, properly grown, could be expected to produce malting samples. H. Hunter, who became Director of the Plant Breeding Institute from 1936-1946, had already bred (with E.H. Beaven in Ireland) the spring malting barley Spratt-Archer, which was the dominant malting barley in Britain until the end of the Second World War. G.D.H. Bell, who followed Hunter, foresaw the value of a winter-hardy malting barley adapted to British conditions. Winter-hardy barleys were already being developed in continental Europe but for general purposes, not for malting. In 1943, Bell released the winter-hardy malting barley Pioneer, followed in 1944 by the variety Prefect and in 1947 by the variety Earl. There was a gap in the breeding programme during the war years; the next advance was the release by Bell in 1953 of the famous two-rowed spring malting barley Proctor, derived from a cross of Spratt-Archer with the high-yielding Danish variety Kenia. This combined high malting quality with a yield 10% above that of Kenia. It quickly came to occupy 70% of the current barley acreage and remained acceptable until the 1970s.

The highest-yielding winter barleys on the best soils probably approach the better yields of wheat (about 10 t ha^{-1}) but much barley, particularly spring-sown, is grown on poorer soils and with restricted fertiliser application for the production of malting barley; it has yields of $3–5 \text{ t ha}^{-1}$. Bell was joined at Cambridge by Sparrow, Roger Whitehouse, Jenkins and Riggs; together or singly they were responsible for the further development of the programme. The next advance was the release of the winter barleys Maris Otter (1963) and Maris Puma (1964). Maris Otter became the most successful winter-hardy barley of malting quality and occupied that particular niche for more than 20 years until it was succeeded by the variety Halcyon (1983), which promised to be equally successful and is still popular at the time of writing.

Nothing in agriculture is ever simple, however. It is true that after the release of Proctor the barley acreage increased for some years more or less in step with the increased acreage devoted to that variety. At the same time winter malting barleys increased, probably contributing to the increased acreage after the influence of Proctor had begun to wane. Other factors were involved in the increase. Winter-hardy feeding barleys from continental sources (none was successfully produced by the U.K. state breeders) became highly popular for farming reasons. Once their reliability and yield were proven they successfully spread the harvest and sowing time for barley on arable farms and eased the labour and machinery profiles. They allowed, too, an early entry for oil seed rape. However, because of the difficulty of finding an early entry point for winter barley itself, the farmer was forced in many cases to follow winter barley with winter barley. On the negative side, the winter crops formed a bridge for plant diseases from season to season and probably contributed to the increase in routine spraying which was a feature of the cereal scene at this time. Secondly, the success of the state sector plant breeders in England (reference will be made to Wales and Scotland later) could not remain unchallenged for long. Private breeders in the U.K. and in Europe quickly followed and sometimes took the lead in the production of new varieties, particularly in the case of barley, which is more amenable to improvement than wheat. A new phenomenon then appeared: the private sector companies devoted largely to the marketing and exploitation of the promising new varieties bred by specialised breeding companies. Since it was, at that time, scarcely possible to exploit state-owned varieties in a commercially rewarding fashion, the entrepreneurs were forced to turn to Europe for promising material; men like Sir Joseph Nickerson instituted a series of trials to determine the value of these varieties and then sold them to farmers with all the skill and resources of a large selling organisation. Sir Joseph, a farmer himself, became fascinated by plant breeding; he said it was better than breeding racehorses, and he set up his own breeding company, Rothwell Plant Breeders, which itself contributed to the sum of cereal varieties. The Plant Variety Rights Act 1964 benefited the private breeders; the consequent setting up of a government agency, the National Seed Development Organisation, to market state varieties should have been an additional stimulus to the state-funded breeders. In fact its policy of maximising its return to the Treasury and limiting its contribution to research funds was possibly discouraging to the state-funded breeders.

One other interesting part of the equation was the breeding of the malting variety Golden Promise by Miln–Masters, a private breeding company of which the two components, separately, played an important part in the early plant breeding scene in Britain. Golden Promise was said to originate from a breeding programme depending on mutations produced by ionising radiations (gamma rays). It had an erectoid habit, which may have contributed to its high yield. It was very susceptible to mildew and was bettered in most of its individual agronomic qualities by other varieties, but its combination of qualities (earliness, high malting quality, stability of yield, and tolerance of adverse weather conditions especially at harvest time) made it very popular in Scotland, where it helped to change the balance of cereals and the farming pattern. There it was the main spring malting barley for many years and remained a favourite with specialist growers of malting barley long after it had been outclassed in the National Variety Trials. Incidentally, it was one of the few varieties to have its period of 'rights' (its patent) extended.

The result of all this breeding, selection and promotion of barley (aided, it must be admitted, by favourable legislation and the fall in the oat acreage) was an increase in the area of land devoted to barley from 898 000 hectares in 1953, the year of the release of Proctor, to 2 411 000 hectares in 1966. It remained at this level until government and E.E.C. policy changed the whole balance of agricultural cropping (see Chapter 13). This increase in acreage was accompanied by an increase in production from about 2 million tonnes in 1950 to more than 6 million tonnes in the 1960s, derived from a wide range of barley varieties (winter feeding, winter malting, spring feeding and spring malting) on a wide variety of soils.

Tests for malting quality

Barley breeding in the initial stages required a limited scientific input. The breeder needed clear objectives within a sensible agricultural framework and the determination to achieve them. However, as traditional methodology reached its limits, new methods based on advancing scientific disciplines began to be used. At an early stage the importance of selection for malting quality combined with high yield demanded a rapid test to enable selections with poor malting quality to be discarded. This was provided by E.T. Whitmore's micromalting test and a mathematical input by L.R. Bishop. It is now used world-wide for screening barley progenies for malting quality. Meanwhile Allison and his colleagues, first at the Scottish Plant Breeding Station and later at the Scottish Crop Research Institute (Camm *et al.*, 1990) were developing an understanding of the relationship between the energy required to mill (milling energy) certain varieties of barley and their capacity to yield a high extract of malt. This is achieved by the use of a modified hammer mill with an integral microprocessor, which calculates the milling energy (in Joules) from the deceleration of a flywheel linked to the hammers when a sample of grain is introduced. The very best malting barleys possess a soft mealy endosperm, which requires less energy for milling. The test is repeatable and very quick: up to 150 samples can be determined in one hour. It bids fair to overtake Whitmore's method once its value has achieved wider appreciation.

Breeding techniques

New breeding techniques were developed using double-haploids, produced by crosses with the wild species *Hordeum bulbosum* and subsequent chromosome elimination. Such double-haploids (dihaploids), because they had two sets of identical chromosomes in each nucleus, were genetically homogeneous and enabled studies to be made on the quantitative inheritance of characters such as yield. This gave an increased flexibility and precision to the breeding programme. Because of the importance of malting of barley, the conversion of the grain carbohydrates to sugar during germination and its subsequent extraction for fermentation, there has been a considerable concentration of work on the anatomy, physiology and biochemistry of the germinating grain. Although practical people have always sought barley with a low nitrogen content when choosing malting samples, there is now clear scientific evidence of an inverse relationship between nitrogen content and quantity of malt extracted (Duffus and Cochrane, 1992). Further work has identified the important enzymes in the germination and malting process. Some of these originate in the aleurone layer (a protein-rich layer internal to the pericarp and external to the starchy endosperm) where gibberellic acid has been shown to be important for their activation (Palmer, 1989).

Enzymes and amino acids

Attempts have been made to influence, by plant breeding, the quantity of enzyme and of protein produced in barley grain. Two examples may suffice. It was long practice to supplement the malt in whisky distilling (particularly in grain distilling) with Canadian (Glacier) and Scandinavian (Akka) six-rowed barleys with a high diastatic power, i.e. rich in β-amylase and capable of synthesising high concentrations of α-amylase, limit dextrinase and α-glucosidase. Workers at the Scottish Plant Breeding Station (now part of the Scottish Crop Research Institute) began a study of the inheritance of these enzymes in malting barley.

On a wider scale, they joined with others in the search for high-lysine barleys. Cereal protein, which is an important part of the diet of domesticated animals, is second-class protein, lacking or containing relatively low concentrations of some essential amino acids. However, the variability of cereals is such that a mixture contains most of what animals require except for lysine, which has to be added as a supplement, either in oil-seed meal or as manufactured lysine. If barley could be bred with the appropriate amino acids, particularly lysine, the feeding process would be greatly simplified and cheapened. In barley a number of high-lysine genes have been found, all of which reduce the yield of grain if only because the endosperm is reduced in volume and the grains tend to be shrunken. However, there seems to be hope that yield can be restored without loss of lysine. Crosses made to incorporate the gene for lysine from Risø 1508 into high-yielding European varieties, restored some yield and produced a grain with 6–6.5 g lysine per 100 g protein. Although this grain tended to have lower digestibility, it nevertheless had an overall value as good as normal barley with supplements. However, the economics of high-lysine barley is still a matter of doubt. This was clearly worked out by Munck (1992) in a fashion that demonstrated the difficulty of matching the scientific idea to the market. His argument was complex, but one part was simple. Oil seeds in Europe are artificially subsidised, so barley mixed with oil-seed residues provides a cheaper balanced diet than lower-yielding barley rich in lysine. Therefore there is no current incentive to use or to continue a breeding programme for high-lysine barleys.

Oats and rye

Oats

Breeding programmes for oats at the state research institutes produced new and improved varieties. However, interest in oats has not moved much beyond the specialist market for porridge oats and, of course, as feed for horses in the leisure industry.

Triticale

There have been few breeding endeavours for the improvement of rye but there are programmes at the Cambridge Plant Breeding Institute and at other institutes throughout the world (particularly in Poland) to cross rye and wheat to produce triticale, a 42-chromosome amphidiploid between the 28-chromosome macaroni wheat (*Triticum turgidum*) and rye (*Secale cereale*) with 14 chromosomes. Experience has shown that this plant carries qualities of hardiness from its rye parent and disease resistance from the complementarity between the disease resistance derived from both parents. Because it combines the high spikelet number of rye with the greater number of fertile flowers and larger grain size of its wheat parent, good yields under less than optimum conditions were expected and in some cases achieved. However, it has suffered teething troubles, with chromosome instability, poor development of endosperm in some grains, and grain sprouting, which, until recently, retarded its use in British agriculture.

Disease control in cereals

The importance of disease control in cereals has already been mentioned. At the beginning of our fifty-year period it was largely achieved by established rotations and by manipulation of the crop environment, including varying sowing dates and rates and reducing winter-proud crops by grazing sheep. There was, too, some measure of elimination of varieties that were particularly prone to diseases, leaving successful varieties that tolerated disease attack; and perhaps disease pressure was less because a number of unrelated varieties was grown, fairly widely distributed, on a smaller acreage and at lower inputs of nitrogen fertilisation.

Then a number of things happened. The Second World War demanded full production; government policy towards agriculture became one of support for all grain grown; higher-yielding varieties appeared; and genes for yellow rust and mildew resistance were incorporated in the breeding programmes. But our understanding of the nature of resistance was still developing, and many of the genes incorporated were of the vertical resistance type, controlled by a single major gene, producing initially a spectacular resistance, but one that was frequently and quickly overcome by changes in the genetic makeup of the pathogen. Mostly this problem was overcome by the production of a succession of new varieties incorporating other resistance genes before the disease became epiphytotic, in Stakman's words 'jumping on the plant pathologist's merry-go-round'. Occasionally catastrophe struck, as in the case of yellow rust on the wheat varieties Heine VII (1955), Rothwell Perdix (1966) and Maris Beacon (1970), all highly promising varieties before the new races of the pathogen appeared and each destroyed commercially in the first season after the emergence of the new race of rust fungus. An improvement in the situation came with the discovery of poly-tunnel and field methods of inoculating with the relevant pathogens, enabling breeders to incorporate a measure of field resistance (horizontal resistance, controlled by polygenes) into the varieties they were selecting. Of course the concomitant development of effective fungicides and spraying methods eased the problem. The problem was not overcome, however, because deteriorating cost–price relationships and environmental concerns militated against the fungicides and have kept the need for genetically based resistance in the breeders' sights. In addition, a reasonable degree of resistance 'protects' against genetic change to resistance to the fungicide and demands a lower level of effectiveness in the fungicide.

An alternative method using multiline varieties, composite varieties or straight mixtures of established varieties has a long history starting with Jensen in 1952, Borlaug in 1958 and developed with some sophistication by Wolfe and his colleagues at Cambridge (1984). The presence of genetically diverse populations of host plants dramatically reduces the spread of epiphytotic diseases and also improves overall yields. Its commercial success is inhibited either by the problems of providing a genetic mixture (for disease resistance) within one 'variety', or of finding an appropriate market for a mixture of two or more varieties. Commercial firms are, however, actively pursuing the possibilities; at the time of writing, mixtures of both wheat and barley can be found in the catalogues.

The development of field tests has been important in discovering those varieties that show 'field resistance' (horizontal resistance, which reduces the rate of disease spread within the crop in the field).

Lodging, which has been a major problem in cereal growing, is of two types. The most dramatic is the collapse of areas of crop under the influence of wind and driving rain, under high fertility conditions and with tall varieties of wheat. The other is caused by *Pseudocercosporella herpotrichoides*, the eyespot fungus, first identified as important by Mary Glynn at Rothamsted and worked on by R.C.F. Macer at Cambridge (1961 a,b). Lupton and Macer (1955) demonstrated that a field test could be carried out by placing infected straw pieces at standard intervals along the rows of developing wheat seedlings. By careful standardisation of the technique it was possible to obtain reproducible results, which demonstrated differences between the susceptibility of varieties to the eyespot pathogen. One of the most surprising results was the demonstration that the important variety Cappelle-Desprez was very resistant to this infection. All later varieties have been tested for resistance, some of it derived from Cappelle crosses and some, as in the case of the variety Rendezvous (1987), from the wild wheat species *Aegilops ventricosa*.

Other crops

The three remaining arable crops, sugar beet, brassicas, and potatoes, illustrate yet more interactions between the scientific plant breeding programme and the great forces of international agriculture.

Sugar beet

Sugar beet has a curious place in British agriculture. It was important in European and North American agriculture in the late 1800s (indeed, it was earlier promoted by Napoleon for strategic purposes) and was produced there relatively cheaply under the high summer insolation of a continental climate. It began to be grown in Britain after the First World War to act as a strategic reserve of sugar in case of further wartime blockade. It is unlikely that it would ever have become a normal part of British agriculture without government interference and encouragement. However, once it was introduced it became a valuable part of arable farming, providing a useful break crop, spreading the labour requirements, providing a reliable source of cash and, for animal enterprises, valuable feed materials in the form of beet pulp and tops. In the 1950s research on the breeding of the crop, funded by the Sugar Beet Corporation, was centred at the Cambridge Plant Breeding Institute but was small in comparison with world effort and slow because of the biennial nature of the crop. Plants were required which showed reduced bolting, epitomised in the Cambridge variety Camkilt, which was quickly overtaken by continental varieties with additional qualities. The other important requirements were increased yield of sugar with lower juice impurities, the production of monogerm fruits and later still the development of male-sterile and polyploid forms to simplify seed production procedures. Although some of the scientific thinking behind these developments benefited from a British input, the main production of varieties was by the major, private, continental European, multinational seed houses concerned with the production and distribution of sugar beet seed. The production of monogerm fruits was a considerable breakthrough. Unimproved sugar beet seed consisted of a fused cluster of individual fruits, which produced a tight cluster of seedlings on germination, difficult and expensive to separate and single. Mechanical treatment for the separation of the seeds was some improvement but the production of monogerm varieties by plant breeding allowed a revolution in the cultivation procedures. Seed could now be sown to a stand and the soil treated with pre- and post-emergence weedkillers to control weeds, thus reducing costs for the early part of the cultivation cycle.

Brassicas

Swedes, turnips, kales and rape were important plants for the provision of forage on many dairy and stock-rearing farms before 1936. Although varieties could be identified, their maintenance depended on very strict isolation and selection because of their out-pollination. Of course the swede was one of the early accidental amphidiploids (in this case a classical allotetraploid in which the chromosomes of a sterile hybrid between two dissimilar parents doubled up to allow pairing and normal reproduction). As a contributor of an enormous tonnage of nutritious material on stock-feeding farms, it was the subject of breeding programmes at the Scottish Plant Breeding Station (now part of the Scottish Crop Research Institute) and at the Welsh Plant Breeding Station, where varieties with improved yields, reduced sensitivity to manganese deficiency and reduced susceptibility to the club root disease (*Plasmodiophora brassicae*) were produced. The marrow-stem kales, which were used for dairy grazing or for feeding fresh as cut material, were tall plants which were difficult to handle with a normal electric fence because they overtopped it. A very interesting programme carried out by K.F. Thompson at Cambridge demonstrated that, as in other crucifers, the genetic mechanism for maintaining out-pollination lay in an interaction between the pollen tube and the style whereby pollen tubes of selfed plants grew slowly and never reached the ovule. This could be overcome by self-pollination in the bud when the style was short. Using this information, Thompson was able to derive a number of inbred lines, which, when combined genetically, produced high-yielding, nutritious material with short stems and improved winter hardiness. To achieve maximum heterosis and to avoid the necessity for frequent bud pollination the variety Maris Kestrel was produced using a cross between two three-way cross, unrelated hybrids. This variety was thought to have added about £1 000 000 to the value of the U.K. kale crop in the 1960s. Other work at the Scottish Crop Research Institute demonstrated the production of, among others, Raphano-brassica allopolyploid hybrids (a cross between a kale or cabbage and a radish in which the chromosome content was artificially doubled to bring about fertility). In yield trials where the fresh mass of the harvested material was measured, these plants showed promise of increase in yield over traditional forage kales but they were before their time because traditional sheep husbandry and folding techniques were unable to demonstrate much increased value from these crops.

Oil seed rape became prominent about half way through our historical period. Britain's entry into the European Economic Community in 1973 was a signal for the expansion of the oil seed rape acreage in the United Kingdom. Thereafter the development of the crop depended on the integration of a complex series of initiatives by plant breeders, agronomists, farmers, merchants and politicians, a veritable template for agricultural integration and development and it is perhaps best dealt with in the appropriate Chapters 8 (pp. 149–53) and 13 (pp. 263–4). Suffice it to say here that the plant breeder was required to breed varieties with increased winter hardiness, with improved oil quality and with reduced content of glucosinolates in the meal. Other problems such as the tendency of the pods to shatter before harvest have still to be addressed; still others have been taken care of by serendipity. Thus genes for resistance to persistent soil herbicides have appeared spontaneously in self-sown oil seed rape in maize fields and the problem of the destruction of plants in the winter by marauding pigeons – previously a major problem and still occasionally troublesome – was overcome by the explosion in winter oil seed rape which took place after the development of winter-hardy varieties and the effective dilution of the pigeon population.

Potatoes

The remaining major arable crop plant in British agriculture in our fifty-year period was the potato. Although the potato plant was imported into Europe in the sixteenth century, it was at first confined to gardens and it was not until the early part of the eighteenth century that it became a farm crop. In the nineteenth century it became an established part of the arable rotation in many parts of Britain. The potatoes of that time were all derived from a limited number of importations and consequently had a limited genetic base. Although plant breeders were active and many splendid varieties were released, the situation could best be described as a group of fanciers juggling the genes of the potato to produce better-looking tubers, which also cooked and tasted better. For reasons that will become clear later as we deal with disease resistance, yield was not a critical matter; all potatoes yielded well if they were protected from the hazards of disease. Agronomists and soil scientists, however, found the potato a sympathetic experimental subject and following their advice the standard varieties of the beginning of our period, King Edward and Majestic, could be grown to yield 9–12 t acre^{-1} (21–30 t ha^{-1}) and more in the best fen soils.

One of the important aspects of the potato had already responded to an inadvertent selection. The first importations of potatoes are thought to have been short-day plants, which continued to grow through the northern summer, only being encouraged to produce tubers and to stop growing when days became short. Hawkes (1967) envisaged a slow selection for long-day plants (which would form tubers in northern summers) during their migration from Italy to north-west Europe. Certainly the early volumes of the Royal Horticultural Society's Journal show Thomas Knight distinguishing and writing about potato varieties which were early, mid-season or late. Most of the breeding material now in use is day-neutral.

The problem for the plant breeder did not lie in the yield, the adaptation to latitude or the agronomic qualities of the potato plant, although these had to be maintained while manipulating other characters.

Breeding control of pests and diseases of potatoes

The problem lay with a number of diseases and pests. The first of these, the wart disease (*Synchytrium endobioticum*), was also overcome by plant breeding serendipity. After the establishment of this very serious soil-borne disease in most of the intensive potato-growing areas of Britain, it was the percipience of an agricultural inspector and the observations of the growers in the north of England (Gough, 1920) that demonstrated the existence of two classes of potato varieties: those affected by the disease and those that were apparently immune. A process of sorting of varieties together with a strict quarantine and statutory control programme brought the disease under control. After the establishment of a simple laboratory test to distinguish between the two genetic types, the breeding of new varieties with immunity to the disease became possible; no new varieties could be marketed thereafter unless they were immune to the disease.

The second problem lay in the virus diseases which, distributed in the main by insect or fungal vectors, particularly in the milder parts of the country, accumulated in the potato stocks because of the plants' dependence on vegetative propagation. Once more the first steps were the identification of the individual viruses contributing to the disease complex, and the institution of a statutory inspection procedure to improve and maintain healthier stocks of potato. That having been accomplished it became possible for George Cockerham at the Scottish Plant Breeding Station to produce varieties such as Craig's Defiance, resistant to the important virus X. Resistance to disease was not, however, the answer for many of the viruses of potatoes. The history of their recognition and of the procedures used to develop healthy stocks is complex and one to which many workers contributed. Suffice it to say that without a resolution of the virus problem modern potato growing in Britain and elsewhere could not have developed. Moreover, the production of seed potatoes would not have become such an important element in Scottish arable farming during the fifty-year period without the fundamental scientific backing along with the statutory control.

The third problem lay in the susceptibility of potatoes to infestation by the cyst eelworms *Globodera rostochiensis* and *G. pallida*, which accumulated where potatoes were grown frequently in the rotation to a point at which they were no longer profitable. The discovery by Howard at the Cambridge Plant Breeding Institute of genes for resistance to this pest and their incorporation in the variety Maris Piper (1966) has played a significant part in the preservation of potato growing in England. The mechanism of the gene that causes the roots of the potato to stimulate the hatching of the eelworms, which cannot then find a compatible host root and thus die, has great utility in reducing the population of eelworms over a period of years.

The fourth problem for potatoes was the potato blight fungus (*Phytophthora infestans*). After the great epiphytotics of the middle to late nineteenth century, fresh importations of genetic material enabled breeders to broaden the genetic base of the plant; some of the new varieties showed some apparent 'field resistance' to the potato blight fungus. However, serious breeding against the disease was only begun in the 1920s by Müller in Berlin, and later at the beginning of our period by W. Black at the Scottish Plant Breeding Station (Robertson, 1991) and by breeders in the Netherlands and U.S.A. Resistance to blight was found in the wild Mexican potato *Solanum demissum*, but crossing with the tetraploid *Solanum tuberosum* proved difficult until a matching tetraploid was constructed from *S. demissum* and *S. rybinii*. The first products from the programme were highly successful but after one year became once more totally susceptible to attack by the fungus. Gradually a set of alleles controlling susceptibility and resistance was identified, and then other sets of alleles at different loci. However, no combination of these genes appeared to be able to hold the blight at bay for more than a season. This was a classical instance of Vanderplank's 'vertical resistance'. An improvement in the situation only occurred when new varieties were sought by crossing with *Solanum andigena* from South America, when a form of horizontal or durable resistance, probably with a wider genetic base, turned up in crosses and showed every level of attack by the fungus on the progeny from nearly complete invasion of the leaf to complete immunity and with all gradations of infection type in between (Wastie, 1991).

All the activity of the period produced potato varieties, including the Pentland and Craigs varieties of W. Black and G. Cockerham of the Scottish Plant Breeding Station and the Maris varieties of H.W. Howard of the Cambridge Plant Breeding Institute, which cropped better (of the order of 10%), had good culinary qualities and were less dramatically susceptible to attack by the potato late blight fungus. Parallel advances in the knowledge of the life history, epidemiology and forecasting of disease and the help of the new range of fungicides allowed the farmer to place these varieties in a cropping system with a greater measure of security. As has been said before, to disentangle all the factors that go to produce increased yields is difficult, but a measure of the sum of the success can be seen in the reduction in the area under potatoes in Britain from 291 000 hectares in 1959 to 203 000 hectares in 1979 without a marked decline in consumption (Riley, 1981).

Pasture plants

Plant breeding has also been applied to pasture plants. The most successful initiative in Britain was by Sir George Stapledon and his colleagues at the Welsh Plant Breeding Institute at Aberystwyth. However, the history of their efforts is so closely tied up with the other developments in agriculture that it is best dealt with in the following chapter.

In the period of the modern agricultural revolution, plant breeding came of age and achieved many notable successes. Many authors have tried to estimate the value of plant breeding in terms of improvements in yield, field characteristics or disease and pest resistance. Perhaps what is written above and in the rest of this book has demonstrated that the benefits from plant breeding cannot be computed in those simple terms. Rather, products interact with a great many other advances in farming and the food industry to provide crops showing stability of yield, increased ease of harvest and utilisation, improved marketability, and (to the consumer) better value for money, improved palatability, etc. In fact they give farmers more control over their businesses and the consumer greater choice of commodity.

Plant breeding world-wide

Plant breeding in Britain was and is part of a world-wide activity, funded both by the state and by private industry. It is dependent on advances made elsewhere for much of its own success. Work in the U.S.A. at the Federal Research Station, originally at Beltsville, Maryland, and at the great State Experiment Stations; similar work in Australia, New Zealand and South Africa; work closer to home, in Europe, in Germany, France, Holland, Italy and in Russia, all contributed to the advance of the subject. Over the early years of the period considered in this book, the British Colonial Service was active, and plant breeders such as Joseph Hutchinson and Michael Arnold in Uganda, Norman Simmonds in the West Indies, Peter Posnette in Ghana, and many others worked successfully to improve the locally important crops. Later the great international research stations devoted to rubber, rice, potatoes, dry-land crops, etc. were established. All contributed to the sum of knowledge; in Britain there was a two-way flow with many of the distinguished breeders returning from other parts of the world to head research institutes or academic departments.

By 1936 simple Mendelian genetics with discrete major genes was understood and acted upon. An awareness of the presence of polygenes and continuous variation, which would respond to selection, developed during our fifty-year period. In terms of disease resistance the concept of vertical (major gene) and horizontal resistance (minor or polygenes) was introduced by Vanderplank in 1963 and is only now well accepted by breeders and pathologists. Breeding methods were changing too. Earlier work involved the transfer of a limited number of characters between refined cultivars from a narrow genetic base (pedigree breeding). Introductions of foreign (wild) genes by a back-crossing programme (introgression) were occasionally undertaken, with consequent prolonged effort to return to the original refined cultivar. A third breeding method, which Simmonds calls incorporation, where a locally adapted population is constantly added to and maintained so that it is good enough to enter the adapted genetic base of the crop concerned, is only now being recognised as a safer and more flexible method (Simmonds, 1993). From such a population, selection can be carried out for specific needs, when required.

The future

What of the future? Such advances have been made in the under-standing of plant genetics in the fifty-year period that the technical procedures of the plant breeder today bear no comparison to the fumblings of the plant breeders in the intervening years. In the few years since 1986 the breeder's capacity to manipulate genetic material has taken another large step, so that at the time of writing there seems to be no limit to the manipulations that might be carried out in the future. The discovery of a range of vectors which can carry genes from one plant to another means that future plant breeding programmes can be closely tailored for the transfer of specific valuable genes. What is not clear yet is how easy it will be to establish and develop varieties with good agronomic and field qualities after the addition of the chosen gene.

This precision means that the programmes of the plant breeder will need to be very carefully designed. To progress, they will need a back-up of physiological and biochemical investigation in the relevant area. Knowledge will be required, too, of the effects of transfer of single genes into an alien genetic environment, as will an understanding of the changes in agronomic and field qualities that follow specific gene transfer. Many of those closely involved with this type of research believe that biotechnology will contribute directly to plant breeding only in a limited way. That is to say, it will not be the immediately chosen route for the production of commercial varieties but its contribution will be in the use of its techniques and tools to transform material, which may then be used in more conventional breeding programmes.

There are three areas where one might expect rapid advance. The problem of plant disease resistance and the recognition of the pathogen by the host is an important area for understanding and, once understood, should enable the tailoring of resistant varieties to be carried out with relative ease. The second area of possible advance is the transfer of genes for the production of important base materials for manufacture or materials for pharmacological or medical use (see Chapter 13). But the value of this approach will depend on a parallel development of manufacturing techniques that can handle the materials produced in the plants. Yet another area of possible advance concerns the adaptation of plants to particular ecological situations. Much of the irrigated area of the world is suffering from incipient salinity, which is inimical to many cultivated plants, yet there are wild plants that grow easily under these conditions. Although there is evidence that salt resistance will yield to selection in non-adapted plants, the possibility of transferring the capacity to grow under saline conditions from already adapted wild plants to cultivated plants is attractive. It will depend on parallel advances in understanding of the physiology and ecology of salt tolerance. Similar possibilities exist for frost- and drought-resistance; indeed, some advance in all these areas is already taking place.

But genetics can not only offer techniques for transfer of specific properties from one plant to another; it can also help to solve problems in physiology and biochemistry by the genetic analysis of complex biochemical and physiological processes. Plant breeding has contributed greatly to the advances in agriculture over the period; its potential for future contribution is enormous.

Selected references and further reading

Austin, R.B. (1987). Physiology department. In *The Plant Breeding Institute, 75 years, 1912-1987*, ed. Joan Green, pp. 53–56. Plant Breeding Institute, Cambridge.

Barnes, P.J. (1989). Wheat in milling and baking. In *Cereal Science and Technology*, ed. G.H. Palmer, pp. 367–412. Aberdeen University Press.

Biffen, R.H. (1905). Mendel's laws of inheritance and wheat breeding. *Journal of Agricultural Science* 1, 4–48.

Biffen, R.H. (1907). Studies in the inheritance of disease resistance. *Journal of Agricultural Science* 2, 109–28.

Camm, J.P., Ellis, R.P., Swanston, J.S. and Thomas, W.T.B. (1990). Malting quality of barley. *Annual Report of the Scottish Crop Research Institute*, Invergowrie, Dundee.

Chollet, R. and Ogden, W.L. (1975). Regulation of photorespiration in C_3 and C_4 species. *Botanical Review* 41, 137–79.

Duffus, C.M. and Cochrane, M.P. (1992). Grain structure and composition. In *Barley: Genetics, Biochemistry, Molecular Biology and Biotechnology*, ed. P.R. Shewry, pp. 291–317. C.A.B. International, Wallingford, Oxford, U.K.

Fowden, L. (1981). Science in crop production. In *Agricultural Research 1931-1981*, ed. G.W. Cooke, pp. 139–59. Agricultural Research Council, London.

Gough, G.C. (1920). Wart disease of potatoes (*Synchytrium endobioticum* Perc.). A study of its history, distribution and the discovery of immunity. *Journal of the Royal Horticultural Society* **45**, 301–12.

Hawkes, J.G. (1983). *The Diversity of Crop Plants.* Harvard University Press, Cambridge, Massachusetts.

Hawkes, J.G. (1967). Masters Memorial Lecture: The history of the potato. *Journal of the Royal Horticultural Society* **92**, 207–24; 249–62; 288–302.

Lupton, F.G.H. (1987). Historical survey. In *The Plant Breeding Institute, 75 Years, 1912-1987*, ed. Joan Green, pp. 6–20. Plant Breeding Institute, Cambridge.

Lupton, F.G.H. (ed.) (1987). *Wheat Breeding: its Scientific Basis.* Chapman and Hall, London and New York.

Lupton, F.G.H. and Macer, R.C.F. (1955). Winter wheats resistant to eyespot. *Agriculture* **62**, 54–6.

Macer, R.C.F. (1961a). Saprophytic colonization of wheat straw by *Cercosporella herpotrichoides* Fron., and other fungi. *Annals of Applied Biology* **49**, 152–64.

Macer, R.C.F. (1961b). The survival of *Cercosporella herpotrichoides* Fron. in wheat straw. *Annals of Applied Biology* **49**, 165–72.

Manners, J.G. (1950). Studies on the physiologic specialisation of yellow rust (*Puccinia glumarum*) in Britain. *Annals of Applied Biology* **37**, 187–214.

Monteith, J.L. (1977). Climate and the efficiency of crop production in Britain. *Philosophical Transactions of the Royal Society of London* B **281**, 277–94.

Munck, L. (1992). The case for high-lysine barley. In *Barley: Genetics, Biochemistry, Molecular Biology and Biotechnology*, ed. P.R. Shewry, pp. 573–601. C.A.B. International, Wallingford, Oxford, U.K.

Palmer, G.H. (1989). Cereals in malting and brewing. In *Cereal Science and Technology*, ed. G.H. Palmer, pp. 61–242. Aberdeen University Press.

Penman, H.L. (1968). The earth's potential. *Science Journal* **4**, 43–7.

Raymond, W.F. (1981). Grassland research. In *Agricultural Research 1931-1981*, ed. G.W. Cooke, pp. 311–23. Agricultural Research Council, London.

Riley, R. (1981). Plant breeding. In *Agricultural Research 1931-1981*, ed. G.W. Cooke, pp. 115–37. Agricultural Research Council, London.

Robertson, N.F. (1991). The challenge of *Phytophthora infestans*. In *Advances in Plant Pathology*, vol. 7, ed. D.S. Ingram and P.H. Williams, pp. 1–30. Academic Press, London.

Silvey, V. (1978). The contribution of new varieties to increasing yield in England and Wales. *Journal of the National Institute of Agricultural Botany* **14**, 367–84.

Simmonds, N.W. (1993). Introgression and incorporation. Strategies for the use of crop genetic resources. *Biological Reviews* **48**, 539–62.

Vanderplank, J.E. (1978). *Genetic and Molecular Basis of Plant Pathogenesis.* Springer-Verlag, Berlin.

Wastie, R.L. (1991). Breeding for resistance. In *Advances in Plant Pathology*, vol. 7, ed. D.S. Ingram and P.H. Williams, pp. 193–224. Academic Press, London.

Watson, J.D. and Crick, F.H.C. (1953). Molecular structure of nucleic acids. *Nature, London* **171**, 737–8.

Wolfe, M., Minchin, P.N. and Barrett, J.A. (1984). Some aspects of the development of heterogeneous cropping. In *Cereal Production*, ed. E.J. Gallagher, pp. 95–104. Butterworths and the Royal Dublin Society, London.

8 Integrations and innovations in crop husbandry

So closely integrated are the various elements in the management of a farm that deviation from the established pattern of husbandry has always been exceptional and not undertaken lightly. Every so often periods of innovative risk-taking have emerged to break through the barriers of conservatism, as in the agricultural revolution of 1700–1860 (see Chapter 1). Another such period we have identified in the modern agricultural revolution (1936-86). In the 1920s, a time of severe agricultural depression, any innovation was driven by desperation. An example was the Hosier Bail, a light movable shed for machine-milking cattle out of doors, saving on the cost of housing, cleaning floors and buildings.

As already indicated, the Second World War and the passing of the Agriculture Act of 1947 (and its later modifications) improved the agricultural situation in the U.K. In the war years maximisation of production had been required to feed the population. The Act of 1947 put in place managed and supported markets designed to encourage sustained production, buffered against the fluctuations of world surpluses. It provided a reasonable income for farmers, tolerable wages for their staff and sufficient profit for the ancillary industries dependent on agriculture. These fiscal measures encouraged innovation because incomes from year to year were predictable and reasonably secure. At the same time incomes were tightly controlled overall and closely associated with traditional cropping and animal production patterns. Thus any innovation that allowed intensification, or which provided a non-traditional product, offered the possibility of increased income. In crop production such innovation meant either the development of new and cheaper methods of cultivation and utilisation of traditional crops, or the development of new crops.

The grass crop falls into the first category on two counts (improvement in its husbandry and improvement in its utilisation); the cereals show an increase in the proportion of winter-sown crops to exploit a longer growing season and changes and improvements in husbandry to match the improved varieties; while oil seed rape and maize are new crops and peas an old crop revived.

144

The grass crop

In grassland husbandry, ley farming, the alternation of a grass ley with arable cropping, was a standard practice in the wetter north and west of the U.K. in the nineteenth century; it was only after many vicissitudes that it became fashionable in much of the rest of Britain. Grass had been the concern of the grazier from earliest days but in a systematic fashion only from the time of the first agricultural revolution (Henderson, 1827). With the introduction of rotations we find a 'seeds' break in the Norfolk four-course rotation, where sown red clover or a mixture of clover and grass was cut for hay, was grazed as aftermath, and finally made an excellent entry for wheat.

In the wetter areas of the country it became customary to extend the seeds break by the substitution of grasses and wild white clover (although seeds of the required quality were not readily available), with the continuation of the grass break for three or more years. The ploughed grassland, providing a source of nitrogen, was the preferred entry for oats. These long leys provided the system of alternate husbandry that was widely popular in Scotland and in limited areas in England. This early enthusiasm for grass, which encompassed the use of grass seeds and clover (rather than relying on the spontaneous growth of the native weed grasses and legumes), and the search for leafier and greener plants came to a temporary end with the agricultural depression of 1870 and scarcely revived until the First World War. By that time many pastures were self-sown and of poor quality except in areas where there was a tradition of permanent pasture for cattle fattening, as in the famous pastures of Leicestershire.

Pioneers of grassland husbandry

However, new pioneers emerged; in the early years of the twentieth century William Somerville, Thomas Middleton and Douglas Gilchrist, all born within a year or two of one another and successively Directors of the Cockle Park Experiment Station of Northumberland County Council and Newcastle University, demonstrated the value of the calcium and phosphate in finely ground basic slag for encouraging clovers on the boulder clays of Northumberland and measured the resulting increase in the weight of lambs. Gilchrist also developed the much simplified seed mixture which, as Cockle Park Mixture, was widely used in grassland husbandry (Pawson, 1960). At the same time R.H. Elliot of Clifton Park in Roxburghshire was developing his ideas on grass in leys, with an admixture of deep-rooted plants as soil improvers. These and others set the scene but the national move forward came from the work of Sir George Stapledon. Working for the Food Production Department of the Ministry of Agriculture during the First World War, he was able to observe at first hand the impoverished condition of much of Britain's grassland. A visionary, he saw the way forward as the development of ley farming and improved strains of grasses and clovers. His enthusiasm and persuasion encouraged Lord Milford to back the setting up of the Welsh Plant Breeding Station at Aberystwyth, with Stapledon as Director, in 1919. With the help of a very small staff, including the plant breeder T.J. Jenkin and the grass expert William Davies, he made considerable progress in producing the leafy, persistent varieties and the agronomic conditions for the development of superior pastures (Stapledon and Davies, 1948). However, uptake was slow. British agriculture was once more in recession, and in 1937–38, with the Second World War on the horizon, Stapledon undertook a survey and found large acreages of tumble-down grassland. As a result he was asked to conduct research on the rehabilitation of these run-down pastures. Three farms, financed by the Ministry of Agriculture, were chosen for the work, which demonstrated the value of sown leys. From these farms developed the Grassland Research Institute at Hurley.

Following Stapledon's work the importance of grassland as a productive contributor to British agriculture was established, and the scene was set for the next move forward. That grass crops responded to nutrient supply was accepted by those who practised drowning of water meadows, where a combination of the increased temperature in the flowing water and the addition of nutrients from the dissolved and suspended material improved earliness and yield. The famous Craigentinny Meadows on the sandy soil to the north-east of Edinburgh, in the mid-eighteenth century, received the town sewage and yielded large crops of grass in several cuts per year, for sale to the city's cow-keepers (Symon, 1959). Later still the Rothamsted Park Grass experiments, begun in 1856 and carried on for many years, revealed the response of grass to nitrogen.

Utilising young grass

At the end of the nineteenth century, Sir David Wilson had found by simple chemical analysis that young grass and clover were likely to be more nutritious than the older material. German workers during the First World War had demonstrated great increases in the yield of pastures treated with nitrogen fertiliser and rotationally grazed. Stemming from this, the nutritional quality of young grass and clover was further investigated by H.E. Woodman at Cambridge and S.J. Watson at Jealott's Hill during the inter-war years. An evangelical campaign on the use of nitrogen by Imperial Chemical Industries Ltd., and intensive work by research institutes and advisory and development services, persuaded British farmers of the value of properly managed grass. After the war W.F. Raymond and his colleagues at the Grassland Research Institute (founded in 1949) made great advances in our further knowledge of herbage quality. The identification of the *D* value (the proportion of digestible matter in the dry matter intake), which could be assessed by animal experiments or by *in vitro* methods, allowed more accurate evaluation of the quality of feeds and, with further work, related the grass crop to the Metabolisable Energy System (Chapter 12).

But it is one thing to use grass during the summer at stocking rates sufficient to keep the sward in a state of perpetual youth; it is quite another to find appropriate feeding material to keep the equivalent number of animals over winter. Provision of winter keep is the central problem of temperate agriculture. Of course crops had been preserved as hay or silage from earliest times (Nash, 1978), but the appreciation of the nutritive value of younger material gave a new stimulus to the development of work on conservation for winter use. Machinery that allowed crimping and rapid tedding of grass improved haymaking procedures with younger grass crops. Baling machines to compress the material for storage prevented loss of leafy material and simplified storage and feeding. Automatic bale-handling machines and the invention of the Big Baler (Chapter 4) further simplified the procedures; the introduction of the American Round Baler brought the handling of hay and straw into the realm of tractor work alone. For a time there was interest in the artificial drying of grass, begun in Germany and developed at Billingham in the 1920s, but cost–price relations prevented the takeoff of this system. At the same time the advantage of silage-making as an alternative to haymaking in the unsettled weather conditions of much of pastoral Britain was recognised. Silage-making was not new, but earlier silages were lacking in digestibility and feeding value (Nash, 1978). With advances in microbiology great strides were made in understanding the ecology of the organisms involved in the production of silage and the biochemistry of the process. With advisory monitoring of *D*-values it was possible to ensure that grass was ensiled when it was most nutritious.

The ensilage process

The first essential for the ensilage process was a supply of sugars in the material to be ensiled on which the bacteria (*Lactobacillus, Streptococcus, Leuconostoc* and *Pediococcus*) throve and produced lactic acid under anaerobic conditions. This criterion was met with difficulty by the youngest and most nutritious material, high in protein, which was also the wettest material. Wilting the mown grass in the swathe reduced the amount of moisture. Picking up the swathes with the new generation of machines, which delivered finely chopped material from the swathe to the trailer, also ensured the release of any sugars in the cell sap, although the addition of sugars or other additives was sometimes necessary. Compaction of the mass in the silo excluded the air, prevented heating and allowed the development of the anaerobic *Lactobacillus* and similar bacteria, which fermented the sugars to lactic acid. Butyric acid production, the result of fermentation by *Clostridia*, which made the silage unpalatable and unpleasant to handle, was prevented by keeping the moisture content down and the carbohydrate content up. Silos were sealed in a number of ways to achieve the anaerobic conditions. On a larger scale, bunker silos, which were not necessarily airtight, were lined with polythene and covered between each day's addition of cut grass. Finally the top sheet was sealed to the side using a variety of procedures. Additives were developed, both chemical and biological, to hasten or add to the acidification process, which ensured the preservation of the ensiled material. The design of silos advanced. On a large scale bunkers were constructed in which the silage would be introduced in a series of inclined planes, compacted with tractors, during and after the day's filling (the Dorset wedge method, invented by Richard Waltham, a farmer, and the Hurley scientists). The ability to make good silage, routinely, encouraged ancillary innovations such as self-feeding by the animals, or easy-feeding systems with special side delivery trailers for filling troughs. Large-scale silage making brought with it the dangers of effluent production, which had to be tackled by a number of engineering measures.

Recent years have seen the development of round bales of firmly rolled, unchopped but wilted material, enclosed in polythene. At first these bales were dropped into large polythene bags but now are enclosed in plastic 'Clingfilm' by a special wrapping machine. The method can make good silage but is expensive. Nevertheless it provides an easy method for the smaller farmer to make silage without the expense of special silos or the danger of silage effluents, and it provides flexibility, both in terms of allowing a small quantity of silage to be made from a limited area of surplus grass and because the bales can be transported and fed in the field.

Additives have been found which speed the ensiling process. These are of different kinds. The early work of Virtanen successfully introduced sulphuric and hydrochloric acids to the crop to be ensiled, but the method was not widely adopted in Britain because of the problems of handling the mineral acids. Alternatively, sugars were added in the form of molasses, or formic or propionic acid were used as additives because they helped the natural population of bacteria to achieve the required pH more rapidly and certainly. These additives were not at first uniformly successful because of the difficulty of ensuring adequate distribution, but following the development of precision-chop harvesters they were used more widely. Naerling in Norway devised a method of applying the formic acid during the process of chopping the wilted grass; the method was taken up and advocated first in Britain by British Petroleum. At a later stage bacterial starter cultures were developed as additives to ensure an initial fermentation of the correct type. Additives are not used by all silage makers but it would be true to say that those making large quantities of high-quality silage for carefully controlled animal enterprises generally favour them. In Britain much of the experimental and development work on silage making was led by Watson (Watson and Nash, 1960), McDonald (1981), Raymond (1981), Wilkinson (Wilkinson & Tayler 1973) and Spedding (1965). Raymond and his colleagues were responsible for elucidating the complex relations between the various processes taking place in silage-making and the palatability and voluntary intake of the resultant material, all of which have to be understood to obtain the best production of animals from the feeding regime. In the economic climate of a managed market, from the end of the Second World War to the 1980s, this remarkable advance in knowledge and farming technique allowed the development of grazing systems and of economical housing systems, brought budgeting and precision to animal enterprises, and allowed the development of in-wintering and out-wintering systems for cattle and sheep. In short, it changed the face of animal production.

The landscape changed too. In May and June, where grass is grown, one by one the fields are shaved of grass, which goes for processing as silage or high-quality hay, leaving behind bleached and desiccated stubbles. Where crops are grown some of the fields change to a brilliant yellow as the winter oil seed rape comes into flower; a once green landscape has become a chequerboard of pale brown, green and gold.

Oil seed rape

Oil seed rape was a minor crop in Britain until 1973, from which time it has increased dramatically. Domestic brassicas comprise *Brassica oleracea* (with a haploid number of chromosomes, $n = 9$, and including the cabbages and kales); *B. niger* ($n = 8$, black mustard); *B. campestris* ($n = 10$, turnip); and their amphidiploid (double diploid) derivatives *B. carinata* ($n = 17$, Ethiopian mustard), *B. juncea* ($n = 18$, oriental mustard) and *B. napus* ($n = 19$, swedes and rapes). There are oil seed varieties of the last four. *B. carinata* is grown locally in Ethiopia; *B. juncea* is grown extensively in the cooler parts of India and Pakistan, and in small quantities in England as a constituent of mustard. The two main oil seed rapes in Europe and North America (mainly Canada) are *B. napus* and *B. campestris*. There are autumn- and spring-sown varieties of both species, with the winter form of *B. campestris* being hardier and its spring form being earlier than the *B. napus* equivalents, giving the farmer a wide range of options. In the U.K. and much of continental Europe, until very recently, the *B. napus* varieties have predominated. The crop has a long history, going back to the sixteenth century in written records and being extensively grown in Europe in the seventeenth and eighteenth centuries for the production of oil for lighting and lubrication. It almost disappeared from Britain and Europe in the nineteenth century, with the dropping of trade barriers, and did not reappear again until the mid- to late twentieth century.

In 1942 trials were begun in Canada to ascertain the possibility of domestic production of oil for marine lubrication. Oil from unimproved varieties with erucic acid as one of the main fatty acids was particularly valued for use in marine engines because of its resistance to water and steam.

With the government guaranteeing a price differential, the acreage of oil seed rape expanded rapidly in Canada, but support was withdrawn in 1949 and the acreage shrank as dramatically. However, Canadian growers had learned how to grow the crop under prairie conditions and were able to produce the seed cheaply enough to leave a margin on the prices then prevailing on the world market. At the same time their plant breeders began to improve the quality of the oil for human consumption. Thereafter output increased and in 1984 was 20% of the world supply. Meanwhile the European Economic Community had been formed in 1957. Against a background of volatility in prices of oil seeds, a large negative balance of oil seed imports, an expanding world population and prosperity in the Community, an agricultural policy was formulated which included the provision of a deficiency payment (paid as a crushers' subsidy) for oil seed rape. The subsidy was sufficiently large to encourage the planting of a substantial quantity of oil seed rape. Britain joined the Community in 1973 and growing oil seed rape immediately became profitable there. According to Bunting (1986) from the 31 000 tonnes produced in 1973, production expanded to 914 000 tonnes in 1984. It is still widely grown, although fiscal policy is changing. A restriction in the amount grown was instituted in 1982–3 and set at 2.15 million tonnes for Europe. In addition support for the crop was changed recently to payment for the area planted (area aid) rather than on the tonnage produced. This has allowed an expansion of the area of spring oil seed rape. The spring-sown crop yields less than the winter variety but it is cheaper to grow and does not have the problems of early autumn sowing. On balance therefore it is now nearly as profitable to grow as the winter variety.

Factors favouring oil seed rape

Three other elements were favourable to the success of the oil seed rape crop. One was the finding that oil seed rape made a suitable break crop for cereals grown continuously, or in very intensive rotations, in the arable areas of Great Britain. The second was the rise in health consciousness in the human population, with a move away from the consumption of animal fats and their replacement by vegetable-based margarines and cooking oils. The third was the intensification of the animal industry and the increased use of the high-protein meals remaining after oil extraction for incorporation into animal feeding stuffs.

The fatty acids of oil seed rape

Soon, however, countervailing elements emerged. The oil of the unimproved varieties contained palmitic, stearic, oleic and linoleic acids (of which oleic and linoleic acids constituted over one third of the total). In this it resembled the oils of other crops, such as soybean and sunflower. In addition oil seed rape oil contained linolenic, eicosenoic and erucic acids, the erucic acid at levels from one third to one half of total oil. Linolenic acid is undesirable because it may oxidise to produce unpleasant smells and tastes in cooking. Erucic acid, a long-chain fatty acid with valuable water-repellent characteristics, is known to have deleterious effects on animal heart muscle and is suspected of being dangerous to human health. Other problems occurred: the meal left after the oils had been expressed contained glucosinolates, which broke down to produce deleterious materials after ingestion by single-stomached animals. In addition the husbandry of the crop was not straightforward, with particular problems at harvest time.

What followed was a beautiful example of good development work, where the impediments to the use of the crop were identified, then removed or reduced by research findings from work directed to that end. The problem of the fatty acid composition of the oil was first tackled by Canadian plant breeders, who selected low-erucic-acid varieties in their spring-sown oil seed rape. The genes controlling erucic acid production provided material for other breeding programmes and allowed the development of low-erucic-acid winter varieties. At the same time Swedish workers reported improvements in the palmitic and oleic acid contents, and German workers reported improvement in linoleic acid content. The scene was set for the production of desirable varieties of oil seed rape.

Glucosinolates

Low glucosinolates in the meal took almost twice as long to achieve. The reasons, it is suggested, were that there was considerable urgency in removing the health threat to humans from erucic acid; the genetic problems in incorporation of the genes for fatty acid production were comparatively simple; and, most importantly, E.E.C. legislation progressively reduced the concentration of erucic acid tolerated for acceptance under intervention (the fail-safe mechanism, which provides what is effectively a guaranteed minimum price). In addition, the problems facing breeders seeking to incorporate the low-glucosinolate genes were more difficult (Thompson and Hughes, 1986). These glucosinolates comprise four alkenyl-based compounds and two indole-based compounds. They are all hydrolysed by an enzyme, myrosinase, to give glucose and sulphates. Whereas, two of the alkenyl glucosinolates also release isothiocyanates and the indole glucosinolates release thiocyanates, the remaining alkenyl glucosinolates, progoitrin and gluconapoleiferin, release oxazolidinethiones, including goitrin. Isothiocyanates and thiocyanates, which act as iodine inhibitors, can be counteracted by feeding iodine, but the goitrin interferes with thyroid physiology and for its alleviation requires the more complex and uneconomical procedure of administering thyroxine. Because the oil seed rape meal is bitter as well as poisonous it can only be fed at low concentrations (7–11%), in mixture, to cattle. Even when the glucosinolate content is reduced by the incorporation of low-glucosinolate genes it still cannot be fed to poultry because another constituent, sinapine, is converted to trimethylamine in the fowl and appears as a fishy taint in the eggs of some brown-egg-laying strains. Similar materials (*S*-methylcysteine sulphoxide) are also found at various concentrations in non-oil seed rapes grown as forage for sheep and have varying pathological effects if fed to excess (Chapter 9).

The source of genetic material for the low-glucosinolate programme was the Polish variety Bronowski. This variety's glucosinolate content is determined by three or more partly recessive genes. All seeds from one plant have the same content of glucosinolates, regardless of the pollen parent. The content is maternally determined but the plants from such seeds segregate for different genes. Of course there are many other characters which the plant breeder must take into account, including resistance to shattering, cold-hardiness, and disease. Shattering of pods is a characteristic of weeds and reflects the early stage in domestication of the oil seed rape. Comparison with cereals suggests that with further domestication it might disappear in the hands of the breeder. One other character, which was selected spontaneously, was resistance to atrazine, a pre-emergence herbicide. Rape plants appeared as weeds in maize treated with the material and were isolated; the resistance genes were used in breeding varieties suitable for growing with herbicide treatment in a weed control programme.

Other aspects of oil seed rape cultivation

Oil seed rape has only been grown on a large scale in Europe and Britain for about 20 years, but in that time it has transformed the landscape. It has also transformed bee-keeping in these areas; winter oil seed rape, flowering in May, forms a source of accessible nectar and the flowers are so plentiful that bees transported to the fields can return a honey income even in broken weather, so much nectar can be gathered in such a short time from such a small area. On the other side of the coin, the nectar collected is high in fructose and glucose and apparently without sucrose; although it makes perfectly acceptable honey, this granulates (forms crystals), causes other honey with which it is mixed to granulate also, and becomes solid in a very short time, making extracting and handling difficult. There is as yet no evidence of a causal relationship between the sugar balance of the honey and its granulating property. However, beekeepers are learning to develop systems which make use of the oil seed rape bounty, to build up stocks of bees and to avoid contamination of honey from other sources.

The agronomic problems of oil seed rape were particularly concerned with its establishment and its harvest. In the United Kingdom, until recently, preferred varieties were autumn-sown, winter-hardy except in exceptional circumstances, but needed to be sown early for good establishment. The crop coordinated well with winter barley, which with its early harvest gave it a suitably early entry. Sowing after mid-September has been found to reduce yield considerably. Fungal diseases and pests had to be overcome by pathological research, spray programmes and breeding for resistance. The varieties Jet Neuf, Bienvenu and Rafal, which were grown extensively in Britain during the expansion of the oil seed rape acreage in the 1980s, were all bred by the French company Ringot (Serasem) and possessed resistance to some of the disease as well as good agronomic qualities and low erucic acid. Later varieties (double low varieties) incorporated low glucosinolate content. The final hazard with oil seed rape was the harvesting of the plants without loss of seed from pod dehiscence (shattering). Two routes were followed: early treatment with a desiccant, or early cutting (allowing the crop to lie in the swathe to ripen before picking up with the combine harvester).

Two other crops appeared during our fifty-year period and contributed to changes in the farming scene: forage maize and peas.

Forage maize

Maize was early recognised by European settlers in America as a valuable producer of grain and with incidental value as forage. Following its importation into Europe it spread rapidly in southern Europe both for grain and fodder production but, although its value was recognised by British agriculturists such as Arthur Young and William Cobbett in the eighteenth and nineteenth centuries, it was never successfully established in Britain on any scale. In Europe interest in growing the crop for grain diminished with the importation of cheap maize from the U.S.A. in the latter half of the nineteenth century, reaching three million tonnes by the end of the century. On the other hand, cultivation of maize for forage continued in Europe and developed in the U.S.A. in step with the spread of grain maize. Techniques for ensiling maize were worked out in France by Goffart and Lecouteaux and in a parallel development ensiled maize became one of the main sources of winter feed for cattle in the U.S.A. It was valued for its high carbohydrate: nitrogen ratio, its ease of handling and of course its very heavy yields. Even in the U.S.A. it was some time before the importance of early-ripening varieties for forage and silage were appreciated. High-dry-matter silage was best made from crops that had produced ears with the grain at the 'dough' stage. These ideas, and new varieties and breeding methods for maize developed in the U.S.A., were picked up in Europe by a few pioneers and by an even smaller number of pioneers in the U.K. According to Bunting (1978), whereas in the U.K. in 1965 there were only 1 000 hectares of forage maize grown, there were 460 000 in France and (West) Germany. By 1977 the area had expanded in the U.K. to 36 000 hectares and in France and (West) Germany to nearly one and a half million hectares.

The problems associated with the establishment of maize in the British farming scene were the problems associated with a crop plant at the limit of its climatic range. The primary concern was to find suitable varieties. The great developments in maize breeding were in the United States where in early years two main populations of maize were noted. The tall-growing, non-tillering 'dent' varieties of the southern states had soft starch in the centre of the grain and hard starch at the sides only, so that on ripening the soft starch shrank, leaving an indentation where there was no hard starch. In contrast the maize of the north-eastern states, stretching up towards the Canadian border, was a 'flint' type with a complete covering of hard starch, which maintained the shape of the grain on ripening. These two types were hybridised to produce the corn-belt maize varieties; the programme was extended to produce a range of other types such as pop-corn, sweet corn and flour corn (with a mealy endosperm). An exciting development was the evolution of a breeding system to produce hybrid maize from the crossing of two inbred types to give hybrid vigour. These hybrid maize varieties were found to give very high yields of forage in the U.S.A. and European breeders in Holland and France emulated their American colleagues, producing flint × dent varieties which, for the first time, allowed maize to compete with other forages in terms of bulk and quality at the northern limit of its range (Bunting, 1978).

When suitable varieties had been identified (such as INRA 200 and INRA 258 from the French research service) it became important to study the climatic factors that governed the success or otherwise of the maize crop. Crops at the limit of their range are very sensitive to slight deficiencies in the growing conditions; it soon became apparent that temperature was very important for the success of the maize crop (Carr and Hough, 1978). The initial stages of germination were critical for the early establishment and future well-being of the crop. Germination was poor at soil temperatures below 10 °C, which was reached in late April to early May in the southern counties of England. Thereafter slight differences in the texture of the soil, the colour of the soil, its aspect and its exposure to wind determined the rate of germination and the growth of the crop over the first six weeks when the meristem was still in the soil (Ludwig, Bunting and Harper, 1957; Ludwig and Harper, 1958). The further growth of the maize after that date was determined by the air temperature. At low air temperatures growth was slow and the plants yellow.

A number of methods of estimating the accumulated temperatures required for maize development have been devised (Carr and Hough, 1978); working from historical meteorological data, they can be used to predict the behaviour of maize in different climatic and topographical situations, with regard to temperature. Areas south of Yorkshire were shown to be able to produce a crop of maize with dry matter of 25–30%, suitable for ensiling. Areas north of this were unlikely to be useful for maize for silage or for grain in any but a limited number of years.

The water supply to maize also determined the success of the crop. Periods of soil water stress at the tenth- to twelfth-leaf stage affected the growth of the grain, and there was a relationship between dry matter yield and the length of time the crop was able to grow without water stress. However, experience in general was that, on friable soils with a good organic matter content and good rooting depth, the maize plant could generally grow through the season without irrigation. In warmer summers and on lighter soils irrigation might be necessary. Many of the requirements for water have been quantified by research, and irrigation can be used with precision.

The problem with a new crop must always be how to fit it into the rotational cycle of the farm. Maize is harvested late, mostly too late to permit the entry of an autumn-sown alternative cereal. For this reason many farmers followed maize with maize until it was desired to re-enter the normal cycle, when a spring cereal or spring sown grass crop was introduced. On the other hand benefit has been obtained from sowing winter rye to precede the maize, provided the winter rye can be used so as not to interfere with the sowing date for maize. In this way the total dry matter tonnage derived from one field is increased and introduces an economy into the system.

The pea crop

Peas (*Pisum sativum*) have been grown for dried pea production and as components of forages for many years, but never on an extended scale. They were also grown as a local market garden crop for seasonal sale as green vegetables; near London they were grown on a field scale for this purpose, hence the use of Kelvedon and Feltham in the names of older varieties of garden peas. Consumption of peas was restricted to the appropriate season until the development of the canning process in 1885, at the same time as the mechanical sheller. In the 1930s research workers in Britain discovered how to reconstitute dried peas for canning as 'processed' peas, and shortly after there was a slow development of quick freezing of field-harvested fresh peas (vining peas), which became gradually more prominent in the larger arable areas. The increased area was accompanied by an advanced technology and marketing strategy for the frozen peas; a proportion of the crop was also canned as fresh peas. The area devoted to vining peas grew from 19 000 hectares in 1955 to 57 000 hectares in 1982, when it ceased to expand. A similar growth pattern occurred in the U.S.A. Over the same period there was a reduction in dried pea production in Britain, from 49 000 hectares in 1955 to 11 000 in 1962. However, following the same reasoning that led to increased plantings of oil seed rape, the E.E.C. encouraged the planting of combining peas, which by 1982 had expanded to occupy 28 000 hectares.

Over the period of the modern agricultural revolution, the pea plant became one of the most thoroughly researched and understood crop plants. Many research workers contributed; only a few can be named here. Today the crop occurs in three main categories: vining peas, the dried, harvest or combining peas, and the forage peas.

Vining peas

Vining peas, in contrast to market garden peas, which are picked several times in a season, are required to mature most of their pods at the same time. The pods are combed off the haulm by the vining machine and shelled in a drum before being unloaded into a tanker, which carries them to the freezing plant in as short a time as possible, in a round-the-clock operation. The peas are harvested at an immature stage; the stage of harvest is critical to the whole operation. It is defined as the point at which the pea sugars are just about to be converted to starch and is determined by examination of the size of the peas, by their taste, and by a measure of their softness (estimated by a tenderometer reading). The tenderometer measures the resistance of a sample of peas to penetration by needles linked to a calibrated measuring scale. Processing must follow rapidly after harvest (some latitude can be allowed if the peas are chilled in the tanker); otherwise, the sugars are converted to starch and render the peas unacceptable. Also involved in harvest is the problem of ensuring a flow of delivery to the factory. Sowing dates and varieties are staggered to ensure this and to ensure the continuous use of the very expensive vining machines and their operators. At one period the processor controlled the whole operation through a team of fieldsmen but there has been a move towards control by the growers themselves, who may even employ the processing facilities of a large cold store to process and store their crops. The technology having been developed, the problems are of quality control, logistics and costs, since the margins are very small.

When the peas reach the factory (cold store) they are inspected, cleaned, washed, and blanched with steam or hot water to destroy enzymes which produce 'off' flavours. The prepared peas are then moved over a free-flow fluidised bed, which ensures that they are frozen separately. They are stored at $-24\,°C$ in bins.

Fresh peas for canning are allowed to remain for two days longer in the field before being harvested. They are then treated similarly until the final operation, when they are filled into cans and sterilised before sealing. In the U.K. such peas are usually coloured with approved blue-green dyes, whereas in the U.S.A. and in France the peas are canned without colouring and look grey or khaki-coloured.

The processing of vining peas is a complex operation, affording opportunity for delay at many stages. Accordingly the growing–processing chain is amongst the best coordinated and most complex of any agricultural crop. Each step has demanded detailed research to determine its security.

Combining peas

The agronomic and plant breeding problems of the vining pea are similar to those of the combining pea, which became the subject of detailed investigation in the 1960s. Potential fresh yield at about $6\ t\ ha^{-1}$ was acceptable but actual yields varied from that; work over the period has concentrated on identifying and correcting the factors that might reduce yield. As in all peas, germination rates vary for reasons such as depth of sowing, soil temperature and soil moisture content. In addition, S. Matthews showed that seed lots have been found with differences in 'seed vigour' such that samples with a reasonable germination percentage under laboratory conditions do not show sufficient vigour to germinate successfully under moderately adverse field conditions (Powell, 1985). In addition, work has concentrated on disease and pest control and on weed control.

Diseases, pests and weeds of peas

Diseases of peas can determine the success of the crop. Many soil-borne diseases persist from year to year. *Fusarium oxysporum* f. *pisi*, sometimes called St John's disease because it shows up on or about 24 June, is important world-wide and was shown to be a major problem in the area around London. Work by Charles Walker at Wisconsin demonstrated a number of races of the organism and found vertical genetic resistance, which has been found satisfactory in breeding programmes. Associated with this there are a number of root rots; these form a root rot – *Fusarium* wilt complex, which appears to become worse in fields cropped with peas (or some other legumes) more than once in five years. Resistance is available to some of the fungal components of the complex, but overall it has to be accepted that rotational control is necessary.

Pests of pea crops are also sporadically important and controllable by a variety of methods and sprays. Potentially the most important for the vining pea crop is the pea moth (*Cydia nigricana*), the larvae of which burrow into the peas in the pod and of which a comparatively small infestation can lead to a whole crop being rejected for human consumption.

At the beginning of our revolutionary period, peas were known as a dirty crop, which could be weeded by cultivation at an early stage but which inevitably became weedy when the haulm grew and inter-row cultivations became impossible. The development of pre-sowing, pre-emergence and post-emergence herbicide materials has greatly improved control in the crop. A measure of uncertainty still prevails, however, because of the interaction between moisture, soil texture and the effectiveness of the herbicide, which sometimes leads to failure in weed control. Work on the selection and testing of possible herbicides has been one of the important functions of the PGRO (Processors and Growers Research Organisation) at Peterborough (Gane, 1985).

There is little doubt that the ability to grow peas without weed contamination allowed breeders to widen their horizons in the search for improved yields in vining peas. They began with an effort to understand the growth pattern of the pea plant, which is intrinsically indeterminate in growth (i.e. keeps on growing) but then, as senescence overtakes the apex, becomes secondarily determinate (stops elongating). Genetic and environmental control of the onset of the determinate phase is important for ensuring that a sufficient number of pods mature at the same time for an easy harvest of a uniform product. More recent work by geneticists and physiologists has sought to influence the position of the pods on the stem, to increase ovule number and size, and to reduce the mass and bulk of the haulm. None of this depended on the efficiency of weed control, but the next step, which was to seek ways of dealing with the problem of lodging of peas, did. A gene for leaflessness, discovered in Finland and Russia, was used at the John Innes Research Institute and allowed the development of varieties with much-branched tendrils only (the leafless pea) and of varieties with much-branched tendrils and leaf-like stipules of various sizes (the semi-leafless peas). Growers were at first reluctant to believe that yield was not reduced in these forms, but gradually they became popular, particularly the semi-leafless forms (Snoad, 1985). The plants had the virtue of supporting one another in the field, although they might lodge eventually. They had also the virtue of providing a well-ventilated plant mass, which cut down losses from fungal rots, etc. The open canopy let in more light and allowed weeds to grow unless a good weed control regime was practised.

Forage peas

The forage pea, which is now often a semi-leafless form, has pink flowers; the haulm contains tannins, which prevent bloat when the material, stored as silage or hay, is fed to ruminant animals.

Although great strides have been made in the development of the pea crop, its exploitation has a long way to go. It is a useful source of soil nitrogen in the rotation, but its contribution in a less intensive agriculture has still to be thoroughly evaluated. Improvements must be made in increasing yield and in overcoming the instability of yield which the crop currently shows. Combining peas as a source of animal food have great potential, but the presence of trypsin inhibitors and other anti-nutritional factors is inconvenient. These factors, being mostly heat-labile, are less of a problem when we consider that the possibility of utilising the protein of the combining pea seed in manufacturing processes for human food is only at the very beginning (Wright, 1985).

The cereals

While the new crops were changing the landscape and modifying the farming system, the cereals, the back-bone of the farming system, were undergoing change of a different kind. Chapter 7 tells how the new varieties were yielding much more, were lodging less, were more winter-hardy (in the case of barley), were avoiding more diseases, and were of more predictable quality for brewing and baking. Building on these advances, and on the stability of yield which modern varieties showed, agronomists were able to explore better ways of growing the crops, with results that have transformed the whole rotational pattern. In this period there was a marked decline in the oat crop, associated with the change from horse to tractor power. The decline in demand for this crop was further exacerbated by the relative lack of success in oat breeding compared with that of wheat or barley. In fact the oat produces a valuable commodity with many uses and may be exploited further in the future.

Cereal agronomy

Improvement in yield

There were three thrusts in this work. The first utilised the improved capacity of the new varieties to yield and pushed it to its agronomic limit. The centrepiece of this work was winter wheat. The question arose as to what demands these new wheats made on the minerals in the soil and on the added fertilisers, particularly nitrogen (Spiertz, de Vos and ten Holte, 1984; Batey, 1976). Cannell and Graham (1979) showed that a 10 t ha^{-1} crop of wheat contained at anthesis (flowering) 250, 30, and 310 kg ha^{-1} of nitrogen, phosphorus and potassium, respectively. Because nitrogen is easily leached from soil by winter rains and phosphorus is quickly locked into the soil complex, interest was aroused in the value of applying fertiliser to the crop at intervals (Tinker, 1979).

This work has added interest today, when great concern is being expressed at the problems caused by the leaching of nitrogen to watercourses and ultimately to the water supply.

Agronomically, Y. Coic in 1960 had demonstrated a value in withholding the normal autumn application of nitrogen and providing it in spring at two points to coincide with tillering and stem extension. These ideas were added to and refined by R. Laloux in 1967. The additions were a low seed rate, sown late to fit in with the sugar beet harvest. The low seed rate encouraged tillering by the plant; a further instalment of nitrogen was applied at flag leaf initiation. The wheats used were comparatively tall; they were treated with growth inhibitors to restrict height.

The Schleswig-Holstein system, which grew up in parallel (Effland, 1981), was similar but more intensive. In particular it recommended an early sowing date, close row spacing and a seed rate designed to achieve more than double the population (500 plants m^{-2}, compared with 200 plants m^{-2} in the Laloux system) although the ear densities obtained were only marginally different in the two systems (Falisse and Bodson, 1984). Other similar systems followed and were accompanied by increases in spray applications for disease control and the development of tram lines, which restricted the tractor wheels to particular areas and prevented consolidation of the whole field surface. All these systems had the virtue of affording the farmers guidelines and targets and of requiring them to take a detailed interest in the growth of the crop; above all they formed a talking point for discussion with other farmers and advisors.

There was no doubt that very high yields were achieved consistently on the best soils by using these systems. Whether they were economic was a more difficult question. The answer in any one year depended on the price (and the price support system) then current and the costs of the inputs. In terms of the investment needed, the profitability did not always exceed that of a less intensive system. Various estimates showed that the variable costs of seed, fertilisers and sprays were about one third of the total costs; most of these costs could not be trimmed without loss of yield. The fixed costs of rent, labour, and machinery made up the rest of the costs and were the main reason for any loss of profitability. High-yield systems thus became a challenge to the skilful manager and entrepreneurial farmer who, by spreading fixed costs over a larger acreage, by stretching the labour force or by cheapening machinery costs with a farm workshop, could make the system profitable (Hubbard, 1984).

Much of what is written here about winter wheat applied also to winter barley, but the situation was different with spring barley. There two different products were sought: heavy-yielding feeding barley, and barley that produced a malting sample. Fertiliser applications designed to exploit the yield potential of the feeding barley were in order. More care had to be taken with the malting barley, where excess nitrogen fertiliser led to poorer samples of malt with loss of the malting premium on which the profitability of the smaller crop depended. With the high yields now obtained from winter wheat and winter barley, there was a shift to more autumn-sown crops, although the introduction of winter barley into the spring barley areas of Scotland was resisted for a time. The fear was that the winter crop created a 'green bridge' to carry disease, mildew in particular, onto the then susceptible spring malting variety. In the event more mildew resistant varieties and better fungicide systems allowed both winter and spring barleys to be grown successfully.

Minimal tillage

The second and related thrust in cereal production was the development of minimal tillage and direct drilling systems (Holmes, 1976; see also Chapter 4, pp. 58–9) which, where they could be achieved satisfactorily, were one possible means of cheapening grain production in the expensive intensive systems already described. Direct drilling first became possible with the discovery of the bipyridylium herbicides such as Paraquat (Chapter 6), which killed weeds and allowed seeds to be sown without interference with the surface tilth. The realisation that crops grew well in the absence of weeds *and* the absence of cultivation caused a search for other mechanisms of minimal cultivation which stirred the surface but did not interfere with the deeper soil structure. They all had the benefit of reducing time spent in preparation of the seed bed with consequent saving in labour and fuel, which in turn helped to make the intensive systems more profitable. Experience and experiment have shown that not all soils are suitable for minimal tillage. Cannell *et al.* (1980) concluded that, in Britain, many clay soils were self-mulching in the upper two to three centimetres, and that the cracking and fissuring achieved during the previous summer, along with earthworm activity, provided pathways for the developing cereal roots. Such soils showed yields at least as good as those from soils ploughed with a mouldboard plough in the traditional way. Silt loams, however, showed a yield reduction until a buildup of organic material on the surface had been achieved. Coarse sands, on the other hand, were believed to be unsuitable because of the compaction they showed under field traffic and their low organic matter. Because of the absence of spaces they restrict root growth unless stirred by the normal ploughing and cultivation techniques. Based on this knowledge and other evidence, Cannell and his associates produced a map of the arable areas of Great Britain with indications of the suitability of different soils and districts for sequential direct drilling. In practice, minimal tillage methods rather than direct drilling are used to increase speed of sowing and reduce costs, as equipment normally required on the farm can be used.

Continuous cropping

The third thrust was continuous cereal cultivation, sometimes referred to as sequential cropping. Of course growing the same crop on the same land for several years was not new. Wetland rice cultivation not only followed rice with rice in succeeding years but even twice and exceptionally three times in the same year, year after year. In the U.S.A., wheat cultivation in some of the prairie areas was almost continuous, and dry-land barley followed dry-land barley with only the intervention of a water-saving fallow. However, it could be argued that there was something special about these situations which allowed sequential cropping with the particular cereal. In Britain and western Europe there was a much greater inbuilt belief in the need for rotation, a belief well borne out in the case of legumes and oil seed rape, which do not maintain yields if sequentially cropped.

The change to continuous cropping took place gradually in Britain and was not prompted by any particular research finding, although its continuance and its justification owed a good deal to the support of research and development work. A great many factors came together to make the change possible. First and foremost, mechanisation (Chapter 4) meant that the traditional rotations, which allowed a spread of work on difficult soils (e.g. the wheat and bean rotation on the heavier East Anglian soils), were no longer necessary when mechanical work could take advantage of weather windows and prepare seed beds in shorter time. The arrival of minimal cultivation techniques also afforded the farmer a flexibility in managing a large acreage of cereals on a difficult soil. There were also savings to be made when the operations on the farm revolved around one major crop area.

Other factors were economic: wheat in particular was a profitable crop. Yet again the soil-borne pests and diseases that affected wheat had been considered to require a rotational separation for control. Later work showed that the take-all fungus (*Gaeumannomyces graminis*), which was expected to build up with time in sequential cropping, did so in the early years of the sequence but settled to a reasonable level, which could be mitigated by cultural inputs, in the later years (Ogilvie and Thorpe, 1963). These authors also considered the importance of the eyespot disease. Resistance to the disease was imported from the variety Cappelle-Desprez; the shorter and stiffer-strawed forms were in any case more able to escape infection. In a similar fashion the root eelworm of wheat, which had been expected to devastate sequential crops, failed to do so and settled to an equilibrium situation (Collingwood, 1963). Once wheat was freed from the restrictions of the rotation, barley came under review and systems of sequential cropping developed for it (Shepherd, 1961).

The picture that emerges is of a liberated husbandry, which left farmers much greater freedom of choice in what they wished to produce on their farms and of how they were going to produce it. This freedom came from a series of small discoveries about the way the crops would respond in situations that would offer more options to farmer–managers. They were quick to take advantage of their new-found opportunity.

The green revolution

It would be wrong to leave this discussion, sharply focused as it has been on the British agricultural scene, without reference to the 'green revolution' which at first seemed set to liberate much of the developing world from hunger. A first-hand account of this can be found in *Seeds of Change* by Lester Brown (1970), who was at the heart of the revolution. The many factors involved in improving the production of cereals in the third world and the role of the many agencies in promoting this work is worthy of encyclopaedic treatment. There is only space here to point to the vital role of the discovery of the dwarfing genes of wheat in Japan and their importation into the U.S.A., where they were incorporated into some local breeding material by Orville Vogel. From there they were incorporated into the wheat breeding programme of the Rockefeller Foundation, led by Dr Norman Borlaug, in Mexico. Because he had a large population of breeding material and because he was attempting to grow two generations of wheat each year alternately at sites 800 miles apart, at Mexico City and near the U.S.A. border, the wheats he produced were widely adapted to different day lengths, temperatures and other environmental conditions. They were first successful in Mexico and then transferred on a cost-free basis to a variety of subtropical countries, where they have proved capable of the same high yields. A similar story is told about rice. The International Rice Research Institute, near Manila, was set up by the Ford and Rockefeller Foundations and headed by Robert Chandler. His team followed the pattern of the wheat work and found dwarf rice material suitable for incorporation, from which they produced the 'miracle rice' IR-8 which, with proper management, is capable of doubling the yield of local rice varieties. The spread of these high-yielding crops and varieties to new countries brought a need for advanced agronomic research, for fertiliser research and provision, and for the introduction of new methods of harvesting and threshing, processing and marketing and pricing. All this has formed a focus and an encouragement for the local agriculturists even if success has not been achieved in all the problem areas. On the negative side, growth of the human population has continued in nearly all these countries. As a result, the significant increases in yields, which should have alleviated hunger, have been eroded by the burgeoning population. Nevertheless the success of these varieties, and the perception of the capacity of agricultural technology to deliver, has lifted spirits and given hope in situations which before were characterised by hopelessness. This optimism has reinforced the will of the richer countries to help the poorer by direct funding (though never enough) and through international agencies such as the Food and Agriculture Organisation of the United Nations, as well as the great philanthropic foundations.

Selected references and further reading

Arthey, D. (1985). Vining peas – processing and marketing. In *The Pea Crop, a Basis for Improvement*, ed. P.D. Hebblethwaite, M.C. Heath and T.C.K. Hawkins, pp.

433–40. Butterworths, London.

Batey, T. (1976). Some effects of nitrogen fertiliser on winter wheat. *Journal of the Science of Food and Agriculture* **27**, 287–97.

Brown, L. (1970). *Seeds of Change. The Green Revolution and Development in the 1970's.* Published for the Overseas Development Council by Praeger Publishers, New York.

Bunting, E.S. (1978). Maize in Europe. In *Forage Maize, Production and Utilisation,* ed. E.S. Bunting, B.F. Pain, R.H. Phipps, J.M. Wilkinson and R.E. Gunn, pp. 1–13. Agricultural Research Council, London.

Bunting, E.S. (1986). Oilseed rape in perspective. In *Oilseed Rape,* ed. D.H. Scarisbrick and R.W. Daniels, pp. 1–31. Collins, London.

Cannell, R.Q., Ellis, F.B., Gales, K., Dennis, C.W. and Prew, R.D. (1980). The growth and yield of winter cereals after direct drilling, shallow cultivation and ploughing on non-calcareous clay soils 1974-1978. *Journal of Agricultural Science, Cambridge* **94**, 345–59.

Cannell, R.Q. and Graham, J.P. (1979). Effects of direct drilling and shallow cultivation on the nutrient content of shoots of winter wheat and spring barley on clay soils during an unusually dry season. *Journal of the Science of Food and Agriculture* **30**, 267–74.

Carr, M.K.V. and Hough, M.N. (1978). The influence of climate on maize production in north west Europe. In *Forage Maize, Production and Utilisation,* ed. E.S. Bunting, B.F. Pain, R.H. Phipps, J.M. Wilkinson and R.E. Gunn, pp. 15–55. Agricultural Research Council, London.

Collingwood, C.A. (1963). Continuous corn growing and cereal root eelworm in the southwest. *N.A.A.S. Quarterly Review* **58**, 70–3.

Effland, H. (1981). Un Système intensif en Schleswig-Holstein. *Perpectives d'Agriculture* **45**, 14–23.

Elliot, J.G. (1973). Reflections on the trend to minimum cultivations. *A.D.A.S. Quarterly Review* **10**, 85–91.

Elliot, R.H. (1908). *The Clifton Park System of Farming.* Simkin, Marshall, Hamilton, Kent & Co., London.

Falisse, A. and Bodson, B. (1984). Development of high input systems of cereal production in Europe. In *Cereal Production,* ed. E.J. Gallagher, pp. 269–84. Royal Dublin Society and Butterworths, London.

Gane, A.J. (1985). The pea crop – agricultural progress, past present and future. In *The Pea Crop, a Basis for Improvement,* ed. P.D. Hebblethwaite, M.C. Heath and T.C.K. Dawkins, pp. 3–15. Butterworths, London.

Heath, M.C. and Hebblethwaite, P.D. (1985). Agronomic problems associated with the pea crop. In *The Pea Crop, a Basis for Improvement,* ed. P.D. Hebblethwaite, M.C. Heath and T.C.K. Dawkins, pp. 19–29. Butterworths, London.

Henderson, A. (1827). *The Practical Grazier.* Oliver and Boyd, Edinburgh.

Holmes, J.C. (1976). Effects of tillage, direct drilling and nitrogen in a long term barley monoculture system. *Annual Report, Edinburgh School of Agriculture,* pp. 104–112. East of Scotland College of Agriculture, West Mains Road, Edinburgh.

Hubbard, K.R. (1984). Cereal production systems – the English experience. In *Cereal Production,* ed. E.J. Gallagher, pp. 285–96. Royal Dublin Society and Butterworths, London.

Lazenby, A. (1988). The grass crop in perspective: selection, plant performance and animal production. In *The Grass Crop. The Physiological Basis of Production*, ed. M.B. Jones and A. Lazenby, pp. 311–60. Chapman & Hall, London.

Leafe, E.L. (1988). Introduction – the history of improved grasslands. In *The Grass Crop. The Physiological Basis of Production*, ed. M.B. Jones and A. Lazenby, pp. 1–23. Chapman & Hall, London.

Ludwig, J.W., Bunting, E.S. and Harper, J.L. (1957). The influence of the environment on seed and seedling mortality. III. The influence of aspect on maize germination. *Journal of Ecology* **45**, 205–24.

Ludwig, J.W. and Harper, J.L. (1958). The influence of the environment on seed and seedling mortality. VIII. The influence of soil colour. *Journal of Ecology* **46**, 381–9.

McDonald, P. (1981). *The Biochemistry of Silage*. John Wiley & Sons, Chichester.

Nash, M.J. (1978). *Crop Conservation and Storage in Cool Temperate Climates*. Pergamon Press, Oxford.

Ogilvie, L. and Thorpe, I.G. (1963). The relation of disease control of successful continuous cereal growing. *N.A.A.S. Quarterly Review* **58**, 65–9.

Pawson, H.C. (1960). *Cockle Park Farm*. Oxford University Press, London.

Powell, A. (1985). Impaired membrane integrity – a fundamental cause of seed-quality difference in peas. In *The Pea Crop, a Basis for Improvement*, ed. P.D. Hebblethwaite, M.C. Heath and T.C.K. Dawkins, pp. 383–93. Butterworths, London.

Raymond, W.F. (1981). Grassland research. In *Agricultural Research 1931-1981*, ed. G.W. Cooke, pp. 311–23. Agricultural Research Council, London.

Shepherd, R.W. (1961). Continuous barley growing. *Agriculture* **68**, 248–50.

Snoad, B. (1985). The need for improved pea crop ideotypes. In *The Pea Crop, a Basis for Improvement*, ed. P.D. Hebblethwaite, M.C. Heath and T.C.K. Dawkins, pp. 31–41. Butterworths, London.

Spedding, C.R.W. (1965). *Sheep Production and Grazing Management*. Baillière, Tyndall & Cox, London.

Spiertz, J.H.J., de Vos, N.M. and ten Holte, L. (1984). The role of nitrogen in yield formation of cereals, especially of winter wheat. In *Cereal Production*, ed. E.J. Gallagher, pp. 249–58. The Royal Dublin Society and Butterworths, London.

Stapledon, R.G. and Davies, W. (1948). *Ley Farming*. Faber and Faber, London.

Symon, J.A. (1959). *Scottish Farming, Past and Present*. Oliver & Boyd, Edinburgh.

Thompson, K.F. and Hughes, W.G. (1986). Breeding and Varieties in Oilseed Rape. In *Oilseed Rape*, ed. D.H. Scarisbrick and W.G. Hughes, pp. 32–82. Collins, London.

Tinker, P.B.H. (1979). Uptake and consumption of nitrogen in relation to agronomic practice. In *Nitrogen Assimilation of Plants*, ed. E.J. Hewitt and C.V. Cutting, pp. 101–22. Academic Press, London.

Watson, S.J. and Nash, M.J. (1960). *The Conservation of Grass and Forage Crops*. Oliver & Boyd, Edinburgh.

Wilkinson, J.M. and Tayler, J.C. (1973). *Beef Production from Grassland*, Butterworths, London.

Wright, D.J. (1985). Combining peas for human consumption. In *The Pea Crop, a Basis for Improvement*, ed. P.D. Hebblethwaite, M.C. Heath and T.C.K. Dawkins, pp. 441–5. Butterworths, London.

9 Hunger in the midst of plenty

Farmers are very observant people. They recognise land on which their livestock fail to thrive and often how animals recover when moved to other pastures or when given alternative types of feed. The reasons farmers suggest for the shortcomings of these unsatisfactory grazings are many, ranging from poverty of soil to the presence of flukes or other parasites or of some plant that might be noxious. The alternative of blaming the animals by classing them as 'poor doers' is rarely considered because it is characteristic for some maladies to continue year after year on the same land irrespective of the source of the grazing stock. Furthermore it is recognised that some of these problems are not limited to individual farms but involve whole districts around them. Had farmers in the past been afforded the opportunity to travel they would no doubt have recognised that they were not alone. They would have seen that vast areas of land all over the world are limited in similar ways and there too animals starve and die in the midst of an apparent plenty.

The discovery of the causes of these serious limits placed on animal production has been the result of an international effort in which British scientists have played no small part and in which, in the best traditions of science, there has been excellent collaboration. A few examples are given here. Some no doubt read like complicated stories of detection; others reveal discovery by serendipity; still others show how information gleaned from investigations far removed from the immediate problem has provided clues and solutions. The work has always been difficult, for experimental studies on farms are never easy to conduct with precision. There have been false starts, guesses that proved wrong, and some notable differences between the views and attitudes of investigators. Nevertheless there has been much success and the work has had far-reaching implications, not only in increasing the productivity of world animal production but also in revealing hitherto unknown biological processes. The effects on British animal production were considerable. The United Kingdom, though small in size, is very varied in its topography, geology, climate and farming systems. The problems on millions of hectares of other continents could also be found on a smaller scale within the British Isles. Furthermore, as the intensity of farming increased, as more was requested of land and crops, so the incidence of these hidden maladies increased. Their control became commonplace and made a major contribution to overall output of animal products.

As with many agricultural problems the origin of these 'hidden hungers' had been investigated in the past and partial solutions achieved. Solutions were not necessarily reached entirely within the 50-year revolutionary period which is being considered here. However, during that period an immense amount of investigational work took place to improve understanding and to refine and extend the preliminary work. In the accounts that follow, some of the early work is described, if only to indicate the continuity of science and how it depends on a cooperation of scientists over time. Furthermore, it often takes a long time before research is translated into a farming reality in terms of new technology, and it has been the application of science that has been responsible for increased productivity.

Phosphorus

Phosphorus deficiency in South Africa

Undoubtedly the first discovery of the cause of a widespread malady of grazing animals was that of the role of phosphorus deficiency in the origin of 'lamsiekte' on the veldt of South Africa in the early 1920s. The same disease was evident on other continents too under a variety of names: 'loin disease' in Texas, 'bulbar paralysis' in Australia and a mild form, 'croitich', in the West Highlands of Scotland. The disease had been known since the late eighteenth century. Cattle exhibited depraved appetites, chewed bones and animal remains, became emaciated and died. Records show that as the years passed the disease became more prevalent. Various causes were postulated without proof – presence of a grass toxin or some unknown poisonous plant – and eventually Sir Arnold Theiler was appointed together with H.H. Green, to undertake investigations. Sir Arnold (1867–1936) was a Swiss, who had been in charge of the veterinary services of the Union of South Africa. He had only one hand. Green was a Scot, who trained in chemistry at Glasgow and in Germany before moving to the veterinary research centre at Onderstepoort. Later he returned to England to establish the first biochemistry department associated with a veterinary laboratory and there continued work on similar field problems as well as setting up diagnostic services for veterinarians. Theiler and Green first showed that the immediate cause of death was the botulinus toxin due to type C *Clostridium botulinum* present in the carcases and bones that were chewed. The question then arose as to why there was this depraved appetite. Suspecting a craving for minerals, their first tests using calcium-rich supplements were not successful in reversing the curious behaviour and consequent botulism, but when cattle were given phosphorus-rich salts there was prevention. Theiler and Green then showed that the herbage in problem areas contained only about half the amount of phosphorus present in normal pastures and that there were marked soil deficiencies of the element. With the vast areas involved, the application of phosphate-rich fertilisers was out of the question at that time; recourse was had to distribution of 'licks', that is, boxes of phosphatic minerals that the animals could indeed lick to obtain the element. Attempts were made to immunise animals against the clostridial organism, but these were not successful. Theiler and Green distinguished the original lamsiekte, that is the botulism, from 'styfsiekte', the primary disease of the skeleton due to phosphorus deficiency. Their work was taken up and confirmed in other parts of the world to release millions of hectares of land to productive use by grazing ruminant animals.

Cobalt

'Vinquish' or pine

It had been known for well over a century that in the Solway Firth area of Scotland, on the Border hills, in wide areas of north-east Scotland and indeed on some of the better soils of arable areas, sheep and to a lesser extent cattle failed to thrive. Horses were unaffected, nor were wild species; the disease was one of ruminant animals. The Gaelic name for the condition, which seemed common to all areas, was vinquish; its English name an expressive one, pine. On affected hill grazings the animals lost appetite, wasted away and died; any lambs produced were weakly and usually succumbed. On the arable land where lambs were fattened they failed to grow despite ample feed. Moving the sheep to unaffected areas resulted in recovery.

The first attempts to find a cure were based on informed guesswork. First, the possibility of some widespread infection was eliminated by microbiological work. Then, because the sheep looked anaemic, 'Parrishes Chemical Food', a rather crude preparation containing phosphates and iron and used for anaemic children, was tried by Lyle Stewart, who was then working at the Animal Diseases Research Association, close to Edinburgh. This preparation and other crude iron salts appeared to work; it was claimed that the cause of pine was indeed iron deficiency. This was quite wrong.

New Zealand and Australian studies

What was the same or a closely similar disease of grazing ruminants had been noted in other parts of the world and given a variety of names: 'bush sickness' on the pumice soils of New Zealand, 'coast disease' on the calcareous sandy dunes of South Australia, and 'Denmark wasting disease' on the new lands being brought into cultivation in Western Australia. Cunningham in New Zealand had followed similar lines to those in Scotland and had found that crude iron preparations would prevent the disease. The most effective was limonite, an iron ore from New South Wales, containing ferric hydroxide.

In the early 1930s two parallel investigations commenced in Australia, one led by Hedley Marston in the south and the other in the west by J.F. Filmer and, later, Eric Underwood, who became the principal investigator involved. Underwood showed that limonite in large doses (50 g) would indeed cure Denmark wasting disease; this provided some respite for the unfortunate farmers faced with such misfortune. However, purer iron salts were without effect and analyses of the bodies of affected sheep showed they were adequate in iron content. The anaemia was not the classic one associated with iron deficiency but rather seemed similar to human pernicious anaemia, the cause of which was unknown. Obviously something other than iron was responsible.

Marston started off on other lines. Perhaps thinking about the work in South Africa, he examined the possible involvement of phosphorus deficiency but found it was not that. Underwood, realising that the limonite contained something effective in prevention of the disease, chemically fractionated it and tested the efficacy of each fraction in curing sick sheep. Since fractions completely free of iron were protective, iron was categorically eliminated. Step by step, different chemical elements were also eliminated until he arrived at a preparation containing nickel, manganese and small amounts of zinc and cobalt. Testing these in pairs showed that nickel and cobalt cured sick sheep; in January 1935, Underwood demonstrated unequivocally that the disease was due to a deficiency of cobalt. Marston was aware of the work in West Australia – indeed, the two groups shared information – but he took an informed short cut to reach the same conclusion that Underwood was to reach but a few weeks later. Marston was aware from work by geochemists that the transition metals would have been leached from the shell formations of the coastal dunes and examined these absent elements in terms of their ability to prevent coastal disease. Cobalt was effective.

Eric Underwood and Hedley Marston were very dissimilar individuals in temperament and background. Underwood had emigrated to Australia at the age of eight to join his father, who had preceded him: his early life was one of privation and he was somewhat reserved. Marston was a native Australian and versed in the social graces. It is sad that after that gap of a few weeks between the two announcements of the discovery of the essentiality of cobalt they failed to establish good relations; there was always acrimony. Certainly publication of this major discovery should have been conjoint between these two men of genius.

Subsequent developments

The Australian work was soon confirmed in Scotland by H.H. Corner and A.M. Smith in the Borders, and by James Stewart and R.L. Mitchell, in

Easter Ross and in the Solway Firth area, showing that cobalt prevented pine. Mapping of the areas followed, as new methods for the analysis of cobalt were devised. The amounts of cobalt required by a sheep each day were minute – less than 0.1 mg – and it was found that 100 g of cobalt sulphate applied to an acre of affected land would protect sheep for more than two years. With the extensive areas of unfenced land affected, applying cobalt to the land was obviously difficult. Marston invented the 'cobalt bullet', a very dense mixture of cobalt oxide and a finely divided ferruginous clay. This could be introduced into the sheep's rumen by mouth to sink to the bottom of the reticulum, where it slowly released sufficient cobalt to meet the animal's needs. His invention was patented and remains the preferred method of treatment of stock. Underwood did little more work on cobalt but, curiously, became involved in another disease similar to the lamsiekte of South Africa: toxic paralysis. This too was a botulism, but did not involve phosphorus deficiency; the depraved appetite in the course of which sheep consumed the dead bodies of rabbits poisoned during the extermination programmes was due to simple starvation in times of appalling drought.

The finding that cobalt was an essential element remained somewhat of a curiosity, though of supreme agricultural importance, until in 1948 E.L. Smith of Newcastle discovered that the extrinsic factor in liver that cured pernicious anaemia in humans was an organic compound containing 4% cobalt. It was designated Vitamin B12 or cobalamin. An immense amount of work followed. The chemical structure of Vitamin B12 was elucidated by Dorothy Hodgkin; its involvement in enzyme systems was discovered by S.P. Mistry and B. Connor Johnson in the United States, and methods were devised for its analysis to aid diagnosis of deficiency of the vitamin in both humans and ruminants. For the underlying deficiency that Underwood and Marston had uncovered was of Vitamin B12; cobalt was only necessary to provide part of the cobalamin molecule, the synthesis taking place through microbial activity within the ruminant stomach. There microbial action led to the synthesis of many other related cobalt-containing compounds with no biological effect; the relatively high requirement of ruminants for cobalt no doubt reflects this fact.

Copper

The 'teart' pastures of Somerset

The first mention of a disease of cattle on the so-called 'teart' or 'tart' pastures of Somerset was in 1850; it was investigated for over a century by a succession of agricultural scientists. The area involved is about 100 000 hectares in extent; there is a similar smaller area in Warwickshire. Within a day, cattle turned on to these pastures develop profuse diarrhoea; the red-brown hair coats of the Red Devon cattle pastured there turn a light yellow; the animals stop growing; and milk yield drops precipitously, with serious economic loss. The land concerned is all on the same geological formation, the Lower Lias, where there is little drift coverage. Local farmers can tell one teart area from another simply by the feel of the land when they walk on it. The clay of the teart pastures has a texture different from that of unaffected areas, which often intersperse affected fields. The early work blamed the presence of purging flax (*Linum cartharticum*) in the local flora, or parasitic infection, or the presence of aperient minerals in the water supply, or the presence of some unknown infectious agent. All these explanations were found wanting.

The beginnings of a solution were made in 1938 when A.W. Ling and W.R. Muir of Long Ashton Research Station of Bristol University used spectrographic methods to examine the herbage and, to their surprise, found considerable amounts of molybdenum present. These were soon confirmed by the spectrographic chemists of I.C.I. (Chemicals) Ltd. (now Zeneca plc) at its research centre at Jealott's Hill. W.S. Ferguson, the chemist in charge of animal work, and S.J. Watson, who later became Professor of Agriculture at the University of Edinburgh, continued the studies by showing that they could induce teart by feeding molybdenum. They recognised the similarity of teart to a disease in Holland known as 'lecksucht' or licking disease. Reclamation of land from peat led to disease in cereals; and dressing this land with town refuse resulted in its prevention. The town waste provided heavy elements. C. Meyer and J. Hudig in 1922 showed that copper was effective in controlling the cereal disease. Aware of this demonstration of the importance of copper deficiency, Professor Bouwe Sjollema tackled the problem of lecksucht in cattle grazing in the peat land; later, in 1938, Professor E. Brouwer investigated the illness and diarrhoea of cattle grazing the Wieringermeerpolder, the land reclaimed from the North Sea. Copper prevented the disease in both instances. There was no indication that molybdenum was involved in either case; the disease was analogous to the disease in cereals and arose from a simple deficiency of copper in the soil.

Recognising the similarity of the signs of disease in these Dutch animals to those in Somerset, Ferguson and Watson dosed their experimental animals with molybdenum, to produce diarrhoea and the teart syndrome experimentally, and then with amounts of copper far greater than their usual nutritional need for this element. These large amounts of copper prevented the teart produced by molybdenum. This experiment provided the first evidence of the adverse effects of imbalance in elemental supply influencing animal health; in this instance an excess of molybdenum preventing the physiological use of copper present in the diet. The work was immediately applied in the teart areas to release land from the bondage of a hitherto unsuspected toxicity.

Swayback

The widespread presence of sporadic outbreaks of paralysis of new-born or young lambs was first noted by Lyle Stewart in the 1920s, although local knowledge indicated that the same malady had been present for a long time. Many different names were given to it, the most common one being 'swayback'. In Derbyshire, where losses in some years were over 90% of all lambs born, the name was 'warfa' and in Wales 'cefn-gwan'. It was soon apparent that swayback – or something very similar – was not limited to the United Kingdom; it was found extensively in South America, India, South Africa, Australia and New Zealand. The pathological changes were considerable, consisting of a massive demyelination of the central nervous system to the extent that in many cases cavities appeared in the brain. These changes enabled a group of pathologists at Cambridge to show by an international exchange of specimens that the diseases described in other parts of the world were indeed the same as that found in Derbyshire.

In investigating the disease the first guesses were wrong. The very large area of Derbyshire where the disease was rife, leading to an average mortality of 15%, had been mined for lead by the Romans, and the land was heavily contaminated with lead. A similar association of lead with the disease was noted elsewhere, particularly in Western Australia where H.W. Bennetts, a veterinarian, and F.E. Chapman, from the Laboratory of the Government Chemist, were investigating 'Gingin rickets', a disease identical with swayback. It was postulated that swayback was due to lead poisoning; trials showed that dosing the sheep with crude ammonium chloride to remove lead from the body resulted in some improvement in the appalling statistics of mortality. Pure ammonium chloride, however, was without effect; some contaminant was suspected as the effective ingredient. Then Chapman's analyses of the pastures at Gingin showed that one element that appeared low in concentration was copper. Copper was tried and proved effective. A team from Cambridge immediately repeated the Australian work and confirmed that copper would prevent swayback in the flocks of Derbyshire. However, there were some anomalies, which continued to be revealed by subsequent studies undertaken by H.H. Green from the Weybridge Veterinary Laboratory. The copper contents of the pastures in Derbyshire were far higher than those in Western Australia, indeed were of such magnitude as to preclude a primary deficiency of the element. J.R.M. Innes and G.D. Shearer, of the Cambridge group, concluded in the early 1940s that some other factor must be present to account for the incidence of swayback. That other factor again proved to be the presence of elevated contents of molybdenum in the herbage, a finding that linked the swayback studies to the contemporaneous ones on the teart lands of Somerset, where copper was effective in reversing teart due to excessive molybdenum in the pastures. The link, although in retrospect it seems obvious, was first made by A.T. Dick and L.B. Bull in Eastern Australia, where sheep were dying on a large scale from *copper poisoning*. Dick and Bull showed that they could prevent copper poisoning by giving molybdenum and sulphur; this work was soon confirmed in the United Kingdom. It was indeed taken much further when C.F. Mills at the Rowett Institute first showed the role of sulphides in the gut in binding copper; and later, in 1975, Dick, Dewey and Gawthorne suggested that molybdenum acted by combining with sulphide to form a complex series of thiomolybdates (compounds of sulphur and molybdenum). These were postulated to bind copper in the gut and to affect the metabolism of the element. Mills and his collaborators, notably I. Bremner, then showed that two compounds were involved, first tetrathiomolybdate, which blocked copper absorption, and second trithiomolybdate, which blocked the utilisation of copper stored in the animals' tissues. Later work in the early 1980s exploited this evidence that the thiomolybdates are the most potent known antagonists of copper by developing their veterinary use to combat copper poisoning in sheep and their medical use to control Wilson's disease in people, in which copper accumulates in the liver.

A curious and sad sequel

In the medical literature of 1947 a report appeared drawing attention to the fact that of the seven research workers in the Cambridge team of scientists who had worked and continued to work in Derbyshire on swayback, four had developed symptoms and signs of multiple sclerosis. All had handled brain and other material from affected sheep; careful study showed that this was the only factor in common. Dr E.I. McDougall, who had been a member of the team, approached all the centres in the world where similar investigations had been made to find that no other workers on swayback had developed a multiple-sclerosis-like disease. In two of the Cambridge workers, multiple areas of demyelination in the brain were found when they died. The probability of four cases of multiple-sclerosis-like disease occurring in such a small group is one in a thousand million. This strange and sad happening is unlikely to be coincidental. The occurrence of an epidemic of multiple sclerosis in the Faeroes in 1978 suggests that there may be some environmental factor involved, possibly related to the husbandry of sheep.

Further copper problems

In the years that followed, particularly in the 1950s and 1960s and indeed continuing to the present, more and more evidence accrued to suggest that copper deficiency or a molybdenum–copper antagonism severely limited animal production on many farms in the United Kingdom. Ruth Allcroft, originally from Tasmania, who was a member of H.H. Green's staff at

Weybridge, discovered Caithness pine in cattle grazing on the peats, a copper-responsive disease accompanied by stilted gait. This disease seemed similar to two others, 'falling disease' of Western Australia, in which the heart muscle is affected, and 'peat scours' of New Zealand, a malady similar to that found by Sjollema and Brouwer in the Netherlands so many years before. Defects in wool production in copper-deficient sheep, in which the fibre lost its normal crimp, were discovered by Marston in the Merino breed but were virtually absent in most British breeds. Simple trial after simple trial revealed that in many (but not all) instances, growth of young cattle was restored to normality by copper supplements; in some of the poorer dairy herds there were responses in milk production. It appeared that the 'copper problems' were much larger even than the considerable agricultural importance of swayback had suggested.

There followed a major attack on ways to identify where the copper-responsive diseases might occur and how to identify individuals at risk. The beginning was the study of the soil and its content of the elements involved. The soil surveys of England and Wales had not included analyses for traces of mineral elements, but that of Scotland had done so. These results were used as a basis for mapping likely areas where problems might arise and slowly the spectrographic chemists at the Macaulay Institute at Aberdeen, using the high expertise of R.L. Mitchell, built up a partial picture of the trace metal status of the more important soils of Scotland. The work was slow and exacting, entailing hard work of sampling hundreds of square miles in all weathers followed by time-consuming and highly sophisticated analytical methodology.

A clever way to overcome the delays in conventional sampling and analysis was devised by Professor J. Wells of the Applied Geochemistry Research Group at Imperial College. It was tested first in Africa and then applied in a ten-week period of the summer of 1969 to map the distribution of trace metals in the whole of England and Wales. The method was very simple. The sediments from streams represent material washed down from the catchment; analysis of them reveals the trace element status of the catchment. The results of this investigation were published in 1978 by Professor Webb, Dr Thornton and their associates, who had been responsible for the work. The maps showed the ubiquity of molybdenum and its high concentration not only in Somerset and Derbyshire but in many other areas where its presence was unexpected.

More direct approaches were then applied. Blood was obtained from cattle all over England and Wales and analysed for copper by the advisory chemists of the Agricultural Development and Advisory Service to show that a large proportion of the animals had what could only be regarded as inadequate amounts of copper in their blood. However, when tests were made with these cattle with low copper in their blood, it was found that, although some did, many did not respond in terms of an increase in their productivity when given copper. This was particularly true of the infertility in herds of apparently low copper status and on farms where growth of young animals was poor. On farms that did respond, molybdenum was invariably present; where there was no response, low blood copper appeared to be due to iron excess or some other factor. This led in the early 1980s to a complete reversal of the accepted hypothesis, that many cases of infertility could be due to copper deficiency aggravated by the presence of molybdenum, iron and sulphur, to one in which the primary cause was a toxicity of molybdenum that could be reversed by copper. This hypothesis that molybdenum in excess can reduce fertility is still under contention; it implies that, although primary copper deficiency caused by a molybdenum excess can undoubtedly occur, molybdenum toxicity *per se* is probably more important under United Kingdom conditions than had been recognised. Either can be eliminated by pasture management. Thus the results of early work on the teart pastures showed that uptake of molybdenum by grasses and clovers could be reduced substantially by maintaining good soil drainage and restricting the use of lime.

Copper and pigs

Immediately after the Second World War, funds were made available to the National Institute for Research in Dairying to build new accommodation for experimental work with pigs. The architects, to protect the steel of the pens from corrosion, sheathed their bases with copper. Raphael Braude, who had been studying in the United Kingdom and was unable to return to his native Poland at the outbreak of hostilities, and who subsequently made a considerable contribution to the nutrition and husbandry of pigs, noticed that the pigs were chewing these copper inserts. Eventually they consumed them completely. Thinking that this represented an inner craving for copper he supplemented the pigs' diet with the small amount of copper known to be commensurate with their nutritional need, to find no beneficial effect. The pigs were not lacking copper. He tried larger amounts and showed that the consumption of as much as 250 parts per million of copper in the diet, almost 100 times the nutritional requirement, resulted in a considerable (up to 10%) increase in the rate at which the pigs increased in body mass (Braude and Ryder, 1973). These results were rapidly taken up, not only in the United Kingdom but throughout Europe.

There were, however, difficulties when these findings were applied in the United States of America. Adding 250 parts per million of copper to the diets of some American pigs resulted in high mortality; in Europe this was not the case. This problem was resolved by N.F. Suttle of the Moredun Institute at Edinburgh, who found that these high intakes of copper resulted in an imbalance between zinc and iron, which could be rectified by increasing the intake of the two elements. The substantially lower contents of absorbable zinc and iron in the usual American pig diets (compared with European ones) increased the risk of the imbalance and of copper toxicity. When attention was given to the concomitant dietary concentrations of zinc and iron, the value of copper was equally apparent in America.

There were, however, problems. The pig could tolerate these large amounts of copper but other stock could not. If feed enriched with copper and destined for pigs was accidentally fed to other species, death could result from copper poisoning. Although pigs stored much of this copper in their livers far greater quantities were excreted in the faeces and this eventually reached farmland. Fears that this copper could constitute a threat to the environment led the European Economic Community to introduce legislation that halved the permissible level of copper added to the diets of pigs to 125 parts per million. This quantity still gives an increase in productivity, albeit at a lower rate.

Selenium

As a preliminary to studies of the amino acid requirements of calves during the late 1940s, K.L. Blaxter and W.A. Wood at the Hannah Institute in west Scotland prepared an artificial diet from which nutrients could be removed at will. When this diet was fed the calves grew very well but, suddenly, they died. Post mortem examination showed the cause to be a massive degeneration of the muscles of the body and the heart. The disease resembled one that had been produced in experimental guinea pigs many years previously as a result of dietary deficiency of vitamin E. The diet of the calves contained vitamin E but giving additional amounts to the animals prevented the disease, which was traced to the presence in the diet of codliver oil. Dr. F. Brown, who was later to undertake distinguished work on the biochemistry of the foot and mouth disease virus, fractionated the oil to show that toxicity resided in the highly unsaturated acids of the oil.

These findings might have remained interesting but of no great agricultural significance save for the fact that codliver oil was being recommended for calf feeding by the advisory services. The Hannah workers started looking to see if the muscle degeneration occurred on farms adopting the recommendation, and indeed found many such animals. A surprising thing then happened. G.A.M. Sharman of the veterinary investigation service found cases of this disease where no codliver oil had been used at all; these cases were widely distributed on farms around the Moray Firth, Easter Ross and were later found in the Lothians of Scotland. Here was an hitherto unsuspected disease: 'enzootic nutritional myopathy', more popularly known as 'white muscle disease' with a death rate of up to 20%. The disease was associated with the traditional wintering of beef cows on diets of turnips and straw and was seen in their calves at birth or shortly after, when they were turned out to grass. Field trials in Inverness-shire showed that the disease could be prevented with large doses of very pure synthetic vitamin E. These large doses were about ten times the normal need for the vitamin. There was no evidence of the disease when cows were given these same traditional diets without supplementary vitamin E in other areas. Clearly, some other factor was involved as well as vitamin E.

The Hannah workers failed to find this factor. The solution came from the most unexpected of sources. During the Second World War, Germany had embarked on a programme of producing ersatz foods for its population. One of these was a particular yeast grown on residues from the timber industry. It was intended to add this to bread, but when fed to rats it caused a necrosis of the liver. This was investigated by Klaus Schwartz at Heidelberg and later at Mainz during the war and shortly after to show that the necrosis could be prevented by three compounds, vitamin E, the amino acid *l*-cysteine and a so-called 'Factor 3' present in wheat bran, whey and other foods. Schwartz continued this work when he moved to the United States. There he fractionated the foods he had identified as containing Factor 3, testing each fraction in turn until, in 1957, he obtained milligram quantities of an impure but potent preparation. This had a characteristic garlic-like smell, which suggested the presence of the element selenium, and so it proved. Schwartz had discovered the essentiality of selenium.

Klaus Schwartz was a remarkable and gifted man. He was a pianist of concert maturity who had devoted himself to medicine. The discovery of the essentiality of selenium was not his only major contribution for, using meticulous procedures to prevent contamination in the course of animal tests, he showed that chromium was also an essential trace element, playing a role in tolerance to glucose; he also uncovered the essentiality of a number of other elements.

Schwartz' findings were immediately taken up in Scotland. A field trial was undertaken to show that both selenium and vitamin E protected calves from the myopathy, either separately or together; similar studies in other parts of the world showed that the comparable myopathy of sheep could also be prevented with minute amounts of selenium. And these amounts were minute: about 50 parts per thousand million of the diet sufficed.

One of the major areas where the work was applied was in New Zealand. There it had been recognised for many years that lambs failed to grow satisfactorily on some of the pumice soils, but studies over many years had failed to ameliorate this 'hoggett ill-thrift'. Selenium was tried and proved effective. At that time in Scotland the analytical methods available to determine selenium were slow and laborious and so the workers from the Hannah Institute used the possibility of a growth response in lambs to identify the extent of selenium deficiency. Five thousand lambs distributed on farms throughout Scotland were given selenium or acted as controls; the results showed that growth responses occurred on the sands derived from the Old Red Sandstone and from certain granites, but not on other soils. The area of

Scotland deficient in selenium appeared from this study to be about a million hectares of the Highlands and about a quarter of a million hectares in the arable areas. Subsequent work identified further large areas of England and Wales where there was the risk of occurrence of these selenium-responsive diseases.

Some sequelae

The involvement of the polyunsaturated acids and of vitamin E in the myopathy were not fortuitous, nor were their effects of a pharmacological nature. Further investigation showed that both vitamin E and selenium were involved in the mechanisms whereby the cell protects itself against oxidative stress. The risk that such an oxidative stress will develop is enhanced if diets rich in polyunsaturated fatty acids react with oxygen. Indeed, the fact that myopathy occurs in affected areas when calves are turned out to early spring grass was shown by C. McMurray of the Veterinary Laboratory in Northern Ireland to be associated with the fact that this grass is rich in polyunsaturated fatty acids. Work on mechanisms of cellular protection against oxidative stress and the roles of vitamin E, selenium and other compounds continues to the present day.

The possibility that, with such a widespread soil deficit of an element required in such minute amounts by animals, there might be evidence of impairment in people living in the areas concerned did not escape medical investigators. Nothing amiss was found in Scotland, nor in New Zealand, where the soil deficit is much greater than anywhere in the United Kingdom. This no doubt reflects the varied diet of people, who consume foods brought from many parts of the world. However, in China, where the sophistication of the food system is considerably less than in the western world, selenium deficiency occurs in the form of 'Keshan disease', in which the changes in the muscles are very similar to those found in cattle.

Problems with feeds

Besides the inadequacies of grazings owing to primary deficiencies of elements essential to animals, animal production can be limited by the toxicities of individual feeds in the diet. Many of these limitations were investigated in the fifty-year period being considered and their causes elucidated. The finding of the causes did not necessarily result in the amelioration of the harm that was done. In most instances all that was possible was to eliminate the toxic material from the diet. In some instances, however, breeding programmes could be embarked upon to reduce the concentration of the toxin concerned. An example of the latter was the effort made, mostly by

Canadian, Swedish and German plant geneticists, to reduce the concentration of erucic acid and of glucosinolates in rape seed to produce the present day 'double low' varieties of oil seed rape (Chapters 7 and 8).

The problems of toxic constituents in plants used as feed for livestock were not, of course, limited to the United Kingdom. Throughout the world there were problems of greater magnitude than those encountered here, notably 'facial eczema' in sheep in New Zealand, which gave rise to liver damage and which was eventually traced to a fungus growing on the herbage; and 'subterranean clover disease' in Australia, in which the clover produced an analogue of a sex hormone, resulting in widespread infertility in sheep that consumed it. Three toxic problems that were investigated in the United Kingdom are described below. One proved to be of very considerable international significance.

The toxicity of bracken

During the years of the Depression of the 1920s and 1930s, hill farming was severely depressed and the hill farmers' constant war against the spread of the bracken fern (*Pteridium aquilinum*) by cutting it was curtailed. Bracken crept down the hillsides, resulting in the loss of about 10 000 hectares of agricultural land a year. Records of estates that had become infested with bracken showed that in a period of 70 years animal output had been reduced to less than half by its encroachment.

Bracken had long been known to be toxic but, happily, most – but not all – grazing animals avoid it or refuse to eat it if they are bedded with it. Possibly with the hope that identification of a toxin might lead to measures to make bracken safe for stock, several scientists commenced investigations and found that bracken produced at least three different conditions in stock, each associated with a different toxin. The first to be distinguished, by P.H. Weswig in 1946 in the United States, was a paralysis in horses that had been described in the nineteenth century. This proved to be due to the presence in bracken of an enzyme (thiaminase), which destroyed vitamin B1. The horses showed all the signs of an equine equivalent of the human disease beriberi, and could be cured by injections of the vitamin. Cattle were not affected because the microbial flora of their rumens synthesises the vitamin in considerable quantities. The second disease was a fatal one in cattle that ate bracken over a period of several weeks; it was characterised by J.M. Naftalin and G.H. Cushnie at the Rowett Institute as an acute depression of bone marrow activity, affecting all the many components responsible for the manufacture of the cells of the blood. Professor W.C. Evans of the Biochemistry Department of the University College of North Wales continued these investigations and attempted to find the specific toxins involved. He showed that they could be regarded as 'radiomimetic' (that is, they mimicked ionising radiation) and constituted a mixture of several compounds, particularly flavonols. No single one of the many compounds he isolated and tested could be regarded as the specific bone marrow toxin.

The third disease associated with bracken has proved the most remarkable. In 1978 Professor W.F.H. Jarrett of the Department of Veterinary Pathology of the Glasgow Veterinary School was investigating the incidence of certain tumours of the gut and urinary bladder of older cattle. He noted that the geographical incidence of these tumours was precisely defined. All occurred on farms with a high infestation of bracken and none on farms where stock had no possible access to it. The incidence of the cancer in cattle was 2.5 to 5.0% each year. Jarrett carried out feeding tests to confirm his field observations and then embarked on studies related to the aetiology of the condition. Cattle commonly have warts on their skin, which are benign and are caused by a papilloma virus (or rather, four distinct papilloma viruses, as Jarrett showed). Examination of over 8000 animals showed that benign tumours – papillomas – were present in the gut of almost a fifth of the animals and that these were at the sites where the malignant tumours were later to be found in animals aged at least six years. Careful tests showed that a particular papilloma virus was invariably present in the alimentary tumours, suggesting that a factor or factors in bracken had been responsible for transforming the papilloma virus and making it malignant. The link with the radiomimetic compounds which Professor Evans had isolated from bracken seemed equally obvious; the acute effects of these were to produce the bone marrow lesion while the chronic one was to transform the common benign virus into a cancer.

The implications of this work were considerable in relation to the aetiology of human cancer and obviously questions arise about human safety in bracken-infested areas. The Ramblers' Association (with perhaps exaggerated caution) is indeed advising hill farmers and ramblers to wear face masks in the autumn of the year, when bracken spores are rife. Questions arise, too, about what should be done to combat this serious hazard to animal health, potential hazard to people and drain on the productive land resource of the country.

Kale toxicity

Considerable work was done both at the Cambridge Plant Breeding Institute and at the Scottish Plant Breeding Institute to improve the value to livestock of the fodder brassicas. A particularly useful cultivar was the marrow-stem kale Maris Kestrel. It was relatively short-stemmed, could be grazed with an electric fence and was high-yielding and leafy. J.D.F. Greenhalgh and J.N. Aitken at the Rowett Institute gave the kale to housed cattle as their sole feed to assess its nutritive value and found that after a few weeks the animals passed haemoglobin in their urine and developed a severe haemolytic anaemia. Possible and plausible causes were investigated to no avail; it was then realised that the kale contained some factor inducing haemolytic anaemia, associated with the presence in the red blood cells of Heinz–Ehlich bodies (spherical bodies consisting of oxidised haemoglobin).

To undertake fractionation studies with cattle weighing 300 kg and eating more than 50 kg kale a day for several weeks was a daunting task; attempts were made to induce the anaemia in a small animal. Rats, guinea pigs, rabbits, mice and hamsters all failed to develop anaemia when fed wholly on kale; the smallest animal to do so was the young goat. In a series of tests using goats, R.H. Smith and C.R.A. Earl found that juice expressed from the kale was toxic; this juice freed of protein and lipid material was also toxic. Displacement chromatography was then employed on a large scale and a fraction, rich in acidic amino acids, produced the anaemia. Analysis of the amino acid content of this fraction showed it to contain two unusual amino acids, one more basic than the other. Two fractions were prepared and the suspect more basic amino acid eliminated. The toxin proved to be *S*-methylcysteine sulphoxide. This was synthesised chemically and when given to goats produced severe anaemia in 5 weeks. Further work showed that although *S*-methylcysteine sulphoxide was the compound in kale producing the anaemia it was not the primary toxin: the compound was fermented in the rumen to produce dimethyldisulphide and methanethiol, and the former was shown to be a potent haemolytic toxin. The lack of response to kale by species with simple digestive tracts no doubt reflects the fact that a ruminal fermentation is necessary to produce the toxin.

Considerable work followed to ascertain whether the toxin occurred in other brassicas – and it did. In addition, field tests related to manurial practices and timing of harvests gave some hope that concentrations of the *S*-methylcysteine sulphoxide could be reduced. The possibilities of breeding for a lower content of the toxin were explored, but in the interim the only solution that could be advanced for use by farmers was to limit intakes of kale.

Turkey X disease

In 1960 more than 100 000 young turkeys on farms in the south and east of England died as a result of what was called 'turkey X disease'. The outbreaks continued, involving ducklings, pheasant and partridge poults; these were described by W.P. Blount of the British Oil and Cake Mills Poultry Trials Farm at Stoke Mandeville. The birds became lethargic and developed lesions of the liver, including proliferation of the epithelial cells of the bile duct.

Microbiological studies and transmission studies ruled out an infective agent; the conclusion reached was that the cause must be some poisonous substance in the birds' feed. Possible contenders were tested using ducklings, which responded with bile-duct epithelial cell proliferation in a matter of a few days, but none of the substances tested was responsible. All the outbreaks of the disease were traced to one London feed plant, but later a mill in another area was also found to be producing the toxic feed. The common denominator was the presence of Brazilian peanut meal used in making the diets at both mills. Chemical studies on the material resulted in an extract that concentrated the toxin 250-fold and eventually K. Sargeant, working closely with Ruth Allcroft and R.B.A. Carnaghan of the Veterinary Laboratory at Weybridge, obtained a crystalline preparation, which was not quite pure but which produced the disease when given in very small amounts. The compound fluoresced in ultraviolet light; the extent of the fluorescence of material isolated from consignments of toxic meal correlated well with the severity of the disease produced. The origin of the toxin, however, remained unknown. In the course of examining the Brazilian peanut meal for fragments of known poisonous plants, P.K.C. Austwick of the Commonwealth Mycological Institute noticed that fungal hyphae were present in cotyledonous material from the affected meal but not in those of non-toxic meal. Sargeant succeeded in growing pure cultures of the fungal species present from a particularly heavily contaminated sample of peanuts from Uganda, identified the fungus as *Aspergillus flavus*, and showed that it produced the toxin, which was then given the name 'aflatoxin'. Aflatoxin was not, however, a single compound. Subsequent work revealed the presence of a series of toxic compounds produced by the mould.

This discovery had enormous implications for both animal and human health. The aflatoxin B_1 must be considered to be one of the most carcinogenic compounds known; forty micrograms given to a rat for ten days induces hepatic carcinoma. Understandably, considerable effort has been exerted to prevent the growth of the mould on feeds and foods, particularly in the humid tropics, where climate encourages its growth, and stringent precautions are taken through chemical monitoring to ensure that imported animal feed and human food is free of contamination.

Selected references and further reading

Bennetts, H.W. and Chapman, F.E. (1937). Copper deficiency in sheep in Western Australia: A preliminary account of the aetiology of enzootic ataxia of lambs and anaemia of ewes. *Australian Veterinary Journal* **13**, 138–49.

Blount, W.P. (1961). Turkey "X" disease. *Turkeys (Journal of the British Turkey Federation)* **9** (2), 52, 55–8, 61–77.

Braude, R. and Ryder, K. (1973). Copper levels in diets for growing pigs. *Journal of Agricultural Science, Cambridge* **80**, 480–93.

Bremner, I. (1987). The involvement of metallothionein in the hepatic metabolism of copper. *Journal of Nutrition* **117**, 19–29.

Bremner, I. and Mills, C.F. (1986). The copper-molybdenum interaction in ruminants: the involvement of thiomolybdates. In *Orphan Diseases and Orphan Drugs*, ed. I.H. Scheinberg and J.M. Walshe, pp. 68–75. University Press, Manchester.

Brouwer, E., Frens, A.M., Reitsma, P. and Kalisvaart, P. (1938). Onderzoekingen over de z.g. diarrhoe-weiden (scouring pastures) in den Wieringermeerpolder. *Rijkslandbouwproefstation, Hoorn Verslagen van landbouwkundige Onderzoekingen* **44**, (4) C, 171–202.

Cooper, M.R. and Johnson, A.W. (1984). *Poisonous Plants in Britain and their Effects on Animals and Man*. Her Majesty's Stationery Office, London.

Corner, H.H. and Smith, A.M. (1938). The influence of cobalt on pine disease in sheep. *Biochemical Journal* **32**, 1800–5.

Cunningham, I.J. (1931). Some biochemical and physiological aspects of copper in animal nutrition. *Biochemical Journal* **25**, 1267–94.

Cunningham, I.J. (1946). Copper deficiency in cattle and sheep on peat lands. *New Zealand Journal of Science and Technology* **27**, 381–96.

Dean, G., McDougall, E.T. and Elian, M. (1985). Multiple Sclerosis in research workers studying swayback in lambs: an updated report. *Journal of Neurology, Neurosurgery and Psychiatry* **48**, 859–65.

Dick, A.T., Dewey, D.W. and Gawthorne, J.M. (1975). Thiomolybdate and the copper-molybdenum interaction in ruminant animals. *Journal of Agricultural Science, Cambridge* **85**, 567–8.

Dunlop, G., Innes, J.R.M., Shearer, G.D. and Wells, H.E. (1939). 'Swayback' studies in North Derbyshire. I. The feeding of copper to pregnant ewes in the control of 'Swayback'. *Journal of Comparative Pathology and Therapeutics* **52**, 259–65.

Ferguson, W.S., Lewis, A.H. and Watson, S.J. (1938). Action of molybdenum in the nutrition of milking cattle. *Nature, London* **141**, 553.

Ferguson, W.S., Lewis, A.H. and Watson, S.J. (1943). The teart pastures of Somerset. *Journal of Agricultural Science, Cambridge* **33**, 44–51.

Goldblatt, L.A. (1969). *Aflatoxin*. Academic Press, New York and London.

Innes, J.R.M. and Shearer, G.D. (1940). 'Swayback', a demyelinating disease of lambs with affinities to Schilder's encephalitis in man. *Journal of Comparative Pathology and Therapeutics* **53**, 2–39.

Lee, H.J. (1956). The influence of copper deficiency on the fleeces of British breeds of sheep. *Journal of Agricultural Science, Cambridge* **47**, 218–24.

Marston, H.R., Lines, E.W., Thomas, R.G. and McDonald, I.W. (1938). Cobalt and copper in ruminant nutrition. *Nature, London* **141**, 398.

Muir, W.R. (1936). The teart pastures of Somerset. *Agricultural Progress* **13**, 53–61.

Schwartz, K. (1960). Factor 3, Selenium and Vitamin E. *Nutritional Reviews* **18**, 193–7.

Scott, M.L. (1973). The selenium dilemma. *Journal of Nutrition* **103**, 803–10.

Sjollema, B. (1933). Kupfermangel als Ursache von Krankheiten bei Pflanzen und Tieren. *Biochemische Zeitschrift* **267**, 151–6.

Suttle, N.F. (1975). The role of organic sulphur in the copper-molybdenum-S interrelationship in ruminant nutrition. *British Journal of Nutrition* **34**, 411–20.

Theiler, A. and Green, H.H. (1932). Aphosphorosis in ruminants. *Nutritional Abstracts and Reviews* **1** (3), 1–27.

Thomas, A.J., Thelfall, G., Humphries, D.J. and Evans, W.C. (1963). Inhibition of erythropoesis in the rat by extracts of Bracken Fern (*Pteridium aquilinum*). *Biochemical Journal* **88**, Proceedings, p. 60.

Underwood, E.J. (1972). *Trace Elements in Human and Animal Nutrition*. Academic Press, New York.

10 Better and more productive animals

In the spread of temperate agriculture throughout the world, British animal breeds have been prominent. The Hereford, Aberdeen Angus, Shorthorn and (to a lesser extent) Galloway type cattle were among the early colonists, and the Jersey cow was the basis of the early dairying in New Zealand. Similarly, among the sheep breeds the Southdown and the Kent or Romney Marsh have become prominent in the Antipodes, and the very important Corriedale is an inbred product of crosses between English longwool breeds and the Merino. The meat-packing industry of South America, introduced by British entrepreneurs, was at first heavily dependent on importations of British beef breeds as sires. But lest this appear a narrow nationalist point of view, mention should be made of the Brown Swiss and the Friesian, which had such an important role in world dairying; the various Indian breeds such as the Sahiwal and the Zebu crosses, which were of importance in warmer climates; and the Longhorn of Spanish (not British) origin, which pioneered in the U.S.A. At the beginning of our fifty-year period, British farmers had a deservedly high reputation as stockbreeders and that part of the industry brought considerable financial reward. This reputation continued and in 1986 the value of exports of live cattle for breeding was still £7 million, a sum augmented by a further £5 million for the export of semen. Their very success, however, had brought with it a conservatism and lack of adaptability which, as we shall see, led to changes in breed structure with the importation of genes and breeds. The conservatism of the breeders sometimes proved a difficulty for the scientists and agriculturists seeking to harness breeding skills to modern genetic thought. Nevertheless sufficient of a symbiosis was created between the science and practice of cattle breeding to ensure the survival of an active breeding industry, and some of the scientific discoveries were seminal for animal breeding world-wide. Some of these advances are described below in relation to the physiology of reproduction and the genetic improvement of animals.

Changes in breed structure

In the 1930s the predominant breed for producing milk in Britain was the Dairy Shorthorn. This was a dual-purpose breed in that the castrated calves could be fattened readily for beef. This breed was accompanied by the more specialised dairy breeds, the Ayrshire and the British Friesian (derived from the Friesian-Holsteins of Holland) together with Channel Island cattle and a few Red Polls. By 1955 the number of Dairy Shorthorns had fallen to a quarter of the national herd in England and Wales and a post-war battle between the Ayrshire and the British Friesian had turned decisively in favour of the Friesian. Fifteen years later in 1970 the Friesian accounted for 76% of the national herd, the Ayrshire for 10% and the Dairy Shorthorn for only 6%. In the ensuing 15 years to the mid-1980s the process of breed change continued, but at a slower rate, to give a virtually complete predominance to the 'black and whites'. The term black and whites is used rather than British Friesians because a second change in breed structure commenced with the importation of the genetically related North American Holstein cattle, first from Canada in the 1960s and then from the United States commencing in the mid-1980s, such that now (1991) the proportion of Holstein genes in the national herd is about 20%.

The changes were made by farmers who realised the financial returns to be made from the higher propensity to milk of first the British Friesian and later the Canadian Holstein. The technical means were at hand to make the change in that Friesian bulls were standing at stud at the artificial insemination centres; and the method farmers employed was to use these bulls on their existing cows to grade up their herds. The dual-purpose aspects of the Dairy Shorthorn were no longer necessary because beef bulls produced good animals for fattening when crossed with the British Friesian; indeed, one reason for importation of the Holstein was that many breeders thought that the British Friesian had pandered to a need to keep a few beef characteristics in its makeup whereas the Holstein was unmistakably a dairy animal.

With beef cattle the major change took place with the introduction of the so-called 'exotics' in the shape of continental European cattle. The first was the Charolais. This breed was imported officially in 1962 under control of a committee of the Agricultural Improvement Council on the understanding that it should be thoroughly tested. Understandably, the Breed Societies were in opposition, thinking that importations of what they called the unimproved draught oxen of Europe might threaten their export trade of British breeds. The trials, many of which were coordinated by W. Longrigg, Senior Livestock Husbandry Advisor of A.D.A.S., showed that the growth rate of crosses with continental breeds was improved by 5% or more and efficiency of food use by up to 10%. Furthermore, there were a few calving difficulties although these had been anticipated. Commercial importations of Charolais cattle followed and so did that of other exotics. Those that survived the very practical test of whether they increased the output of beef were few and included particularly the Simmental and Limousin breeds. Each new breed became associated with a breed society, and demand for these cattle was high. Supply was initially very short; prices for pure-bred exotics rose to incredible heights before the market settled down.

The change in the distribution of sheep breeds in Britain in the 50 years from the 1930s to 1980s followed the same general pattern as had that of cattle breeds, with an internal redistribution of existing breeds and some importations, but in addition attempts were made to establish new breeds. The hill breeds remained very stable but lowland ones did not. Initially pure-bred flocks were replaced by halfbreds and mules and there was increased interest in breeds such as the Clun Forest. Importations included that of the Finnish Landrace by the Animal Breeding Research Organisation in 1962. This breed produces large litters; a further advantage is that the second and third lambs delivered are not grossly reduced in birth weight. These attributes were to prove useful in the development of intensive systems of sheep husbandry.

Preferred rams for crossing to produce lamb in the 1930s were the Cotswold or the Suffolks of the eastern counties and the Southdown or Hampshire Down in the more southerly regions. The Suffolk and Oxford Down were also used widely as crossing rams on cross-bred sheep. Introductions of European breeds to compensate for unsatisfactory qualities of the home breeds (including excessive fat content) were all made by farmers, singly or in groups, who had carefully studied the potential of the breeds concerned. The Texel breed was first introduced into Eire by the Eire Department of Agriculture in 1964 and then introduced into the United Kingdom in 1974 by a farmers' group led by I. Johnson of Biggar. This was followed by J. Barber's introduction of the Charolais sheep in 1977 and by M. Glixton's (Fosse Livestock of Suffolk) introduction of the Bleu de Maine. New breeds were also established using classical methods of breeding in a closed flock from a cross-bred base, notably by Oscar Colburn, who produced the Colbred in the mid fifties and by Henry Fell, who produced the Meatline and innovative production systems. An interesting new breed was the Cambridge. This breed, developed for high prolificacy, was registered as a recognised breed in the early 1980s. It was developed by Professor J.B. Owen, Professor of Agriculture at The University College of North Wales, Bangor, when he worked at the School of Agriculture at Cambridge in 1964. He selected as his foundation stock 54 ewes from 11 different breeds (including Lleyn sheep from the Lleyn peninsula in North Wales) that had shown multiple births. They were crossed with the Finnish Landrace and then by back-crossing the content of Finn genes was reduced to 20–25%. Selection followed to establish the breed, which has retained its prolificacy. The possibility of extending this sort of improvement breeding continues, for there remain many as yet poorly evaluated local breeds of British sheep.

With pigs, change in the breed structure has also occurred. Initially the number of breeds in use in the 1930s was reduced, particularly by discarding the coloured breeds – the Tamworth, the Large Black and the spotted pigs –

since these were disliked by the bacon manufacturers. In addition, although they had a reputation for good mothering ability, the saddlebacks, which were both dark and too fat, had to go, although their genes probably remain important in some hybrid breeding lines. Many attempts were made to import the superlative Danish Landrace, the bacon of which dominated the British market. Eventually Landrace pigs were imported from Sweden, rendering the industry largely dependent on two breeds only, the Large White and the Landrace. However, a number of breeding companies were set up to provide superior stock selected for their low fat content, and these pigs found enthusiastic buyers. New breeds were also imported from the United States – Durocs and Hampshires – for use by these breeding companies, who marketed their stock by trade name.

The growth of the poultry industry, described in Chapter 12, was based on hybrids and separation into broiler and layer stocks. As a result the conventional breeds of the 1930s disappeared from farms. Indeed for all classes of stock there was a diminution in the number of viable flocks of some breeds and concern was expressed about the possible diminution in the gene stock that their demise would represent. The Rare Breeds Survival Trust, however, with a growing band of enthusiasts, undertook the task of ensuring that these rare breeds did in fact survive, to the considerable pleasure of many people.

The new animals

In the late 1960s it was quite clear that the hill sheep industry was in a difficult situation. Without heavy government subsidy it could not survive, and every economy had been made in the management of the hill ewe stocks. Traditional shepherding had been almost totally withdrawn, and gatherings limited to the absolute minimal number. After many difficulties in finding funds, a group from the Rowett Institute (K.L. Blaxter, R.N.B. Kay and G.M.A. Sharman) and one from the Hill Farming Research Organisation (J.M.M. Cunningham, J. Eadie and W.J. Hamilton) combined to investigate whether the wild red deer of the Scottish hills and deer forests could be farmed. This animal (originally a forest dweller) had survived on the hills, particularly of Scotland, for some hundreds of years and it seemed possible that it could be the basis of a new industry. The objectives to be achieved were laid down at the very beginning. They were the creation of an industry in which the animals were herded according to the best principles of animal welfare, humanely slaughtered in an abattoir, their meat inspected and every hygienic precaution taken in its subsequent handling and distribution. These objectives immediately excluded the alternative of some sort of ranching operation similar to that adopted in deer parks; what was required was a farming system.

The work began with the capture in the wild of red deer calves, hand-rearing them with a bottle using a ewe milk substitute, and then pasturing them on hill land with some supplementary feed. These animals were obviously tame and could be handled with ease; the critical question was whether the second and subsequent generation could be managed. This, however, presented no difficulties. The final outcome of the work was that the animal could indeed be farmed. It had distinct advantages over the sheep in terms of output of meat per hectare and the yield of high-quality meat with favourable fat content from the individual carcase was about 50% greater than that from sheep. No major disease problems were encountered that could not be dealt with by preventive measures. It was also shown that the inclusion of upland grazing in the system would increase output and later work extended the findings to good land in the lowlands. Commercial development followed, the first deer farmer being Dr John Fletcher of Auchtermuchty, Fife. The deer farmers established The British Deer Farmers' Association, published a journal and also established a marketing organisation. An entirely new industry had indeed been established.

Scattered in the remote areas of mountainous Britain and on some islands are small herds of feral goats, presumably established from animals that escaped centuries ago. The Hill Farming Research Organisation noted that these animals kept on hill land in some ways complemented sheep in the use of pasture species; they consumed rushes, *Nardus* grass, and other coarse herbage that sheep refused. A.J.F. Russel of the Organisation realised, too, that the undercoat of these feral goats was extremely fine and was of cashmere quality. He established a herd of about 200 animals, and examined the production of this valuable fibre. Yield of a cashmere-type fibre from feral goats, however, was extremely low, and attempts to increase it by crossing the feral animals with domestic goats led to a thickening of the fibre and a loss of quality. A search was then made for goats with the potential to increase production of the fibre; goats with this propensity were imported from Iceland, Tasmania and Siberia and crossed with the ferals. The yield of fibre was increased without loss of quality, and a cooperative was set up with a small number of farmers to exploit the findings. This industry is at present minute, but indicates that new departures and adoption of species not normally regarded as of farming significance have a part to play in agricultural progress. Towards the end of the period interest developed in wild boar for meat and in llamas and other camelids for fibre and possibly for meat.

Reproductive physiology

The genetic improvement of livestock has been extremely dependent on the growth of information about the reproductive physiology of farm species. Knowledge about the nature of seasonal breeding, oestrus detection, pregnancy diagnosis, artificial insemination, multiple ovulation and embryo transfer have all been of considerable importance in determining the course of genetic programmes designed to increase the performance of animals in future generations. These genetic implications are given prominence here, but not to the exclusion of farming developments that were wholly dependent on exploiting new knowledge of reproductive physiology.

Frequent lambing

The sheep is a seasonal breeder, ovulating in the late summer and autumn to produce lambs in early to late spring. Throughout the summer the ewe is in anoestrus. A major, desirable attribute of a ewe is that she should produce a large number of viable lambs per year so as to spread the cost of maintaining her for that time. One way of achieving this result (which will be discussed later) is to breed to increase the size of her litter at each conventional lambing; another is to break the constraint of seasonal lambing and to mate her more than once a year. The immediate determinant of sexual activity in seasonally breeding animals is the photoperiod. This was first shown by W. Rowan in Canada in birds in the 1920s. In sheep, lengthening nights in the autumn trigger, through a series of neural and endocrine responses, secretion of gonadotrophic hormones by the anterior pituitary. These lead to enlargement of the testis in the male and concomitant secretion of testosterone and in the female behavioural oestrus and ovulation. Artificial simulation of the autumn light pattern does the same.

Manipulation of the light environment was used by J.J. Robinson at the Rowett Research Institute to devise a system of sheep production on a scale commensurate with practical application; he used it to elucidate the nutritional needs of the pregnant and lactating ewe and the management problems involved in frequent lambing.

The ewes Robinson employed were crosses between the Polled Dorset (for its tendency to have a naturally long breeding season) and the Finnish Landrace (for its prolificacy). These were housed in buildings that were darkened and in which light could be controlled. Under decreasing day-length patterns of illumination, the ewes were then synchronised in oestrus so that they all came into season at the same time. This was accomplished by the use of progesterone, the hormone produced by the corpus luteum of pregnancy. R.H. Dutt and L.E. Casida in the United States had found in 1948 that sheep would return to oestrus approximately two days after the last of a series of injections of progesterone; in 1965 T.J. Robinson in Australia devised a progesterone-impregnated sponge which, when inserted into the vagina, provided sufficient progesterone to do the same. J.J. Robinson's sheep were treated with the sponges and were then served by the ram or artificially inseminated; pregnancy ensued. After parturition, lactation lasted one month only. The lambs were weaned and transferred to diets based on barley. (These diets and their effects are described in Chapter 12.) The light regimen was then changed to give a further period of declining day lengths and the cycle, which lasted seven months, was repeated. Over five successive pregnancies the mean annual output of lambs reached 3.5. Remarkably, it was then found that control of day length was not necessary: the induction of oestrus through progestagens (hormones of progesterone type) was sufficient. Most importantly, the remarkable ability of the Finn–Dorset to sustain oestrus cyclicity, without hormone treatment, well beyond the time when other breeds revert to anoestrus in the spring, eliminated the need for blacked-out buildings. This resulted in the adoption of the more frequent breeding technique, first in Northern Ireland, later in Scotland and then world-wide.

The synchronisation of oestrus in all types of stock is important, not only for out-of-season breeding, but also for artificial insemination, embryo transfer and as a management tool. There are great advantages in having parturition take place over a concentrated period of time.

A test for pregnancy

Whether by natural service or by artificial insemination, the incidence of pregnancy in herds of cows is never 100%. Some 25% of cows fail to conceive, and some may conceive only to suffer from loss of the conceptus several weeks later. It is obviously important to identify the pregnant individuals. Progesterone, the hormone of pregnancy, is also produced during the first 14 days of the oestrus cycle after which, if the cow is not pregnant, its concentration in the blood falls. If the cow is pregnant, blood concentrations of progesterone are maintained. That this phenomenon could serve as a basis for a test for pregnancy was obvious: a determination of progesterone in blood is hardly a practical one for farm use. R.B. Heap, who later became Director of the Institute of Animal Physiology and Genetics of the Agricultural and Food Research Council, showed experimentally that progesterone was secreted in milk, and devised a sensitive test based on monoclonal antibodies to determine nanogram quantities of progesterone in milk. This assay was tested in collaboration with J.A. Laing of the Royal Veterinary College and found to be very effective. It was adopted by commercial firms and by the Milk Marketing Board for routine use. In the first six months of the latter's scheme for pregnancy diagnosis, the test was used to test 40 000 cows. The test is now used around the world.

Artificial insemination

On occasion Nature can be incredibly prolific. A single ejaculate of about 4 ml of semen from a bull contains about one and a half billion spermatozoa, sufficient, with present techniques, to inseminate about 2000 cows. The ram produces about 1 ml of a denser suspension of spermatozoa than the bull, while the boar produces about 120 ml with a lower sperm density.

The pioneer of artificial insemination was the Russian physiologist E.I. Ivanov. He used it in 1899 to inseminate mares, using sperm suspended in an artificial medium, and then in 1919 the Soviet Government established a centre to inseminate both mares and ewes to replace stocks depleted by the war. In the United Kingdom the work was taken up by the Cambridge School under John Hammond and particularly by Arthur Walton. Some famous experiments were done, such as posting semen to Edinburgh to demonstrate that sperm could be transported and result in pregnancy – rather than to show the reliability of the postal service at the time!

For collection of semen the bull was allowed to mount a cow in oestrus and its penis was diverted to an artificial vagina. Separation of the accessory fluids was made and the semen was then diluted. Considerable work was undertaken to arrive at a satisfactory diluent, which was egg yolk with citrate buffer.

In 1942, in the depths of the Second World War, the first artificial insemination centre was set up at Cambridge under the Agricultural Improvement Council of the Ministry of Agriculture, with L.E.A. Rowson in charge; a further centre was then opened at Reading. The attraction of artificial insemination at that time related to its cost relative to natural service as well as to the reduction in the risk of disease transmission. Many dairies at that time had fewer than 10 cows, and the expense of keeping a bull was several times greater than the cost of artificial insemination for such small numbers. Some small farmers borrowed a bull, with the risk of disease transmission. The Breed Societies were worried about maintaining records of parentage for their herdbooks. Initially their representatives attended the collection of semen, its transport to farms and the insemination of the pedigreed cow, and often the birth. This did not last for long!

Preservation of sperm by freezing

It had been found that cattle spermatozoa could be kept at about 4 °C for a few days. What was required was a technique for keeping them longer. In the early 1940s A.S. Parkes, working at the National Institute of Medical Research, had commenced a programme of work on cryobiology and had appointed C. Polge and Audrey Smith to investigate the possibility of preservation of fowl spermatozoa by freezing. After many abortive attempts, a chance observation showed that fowl spermatozoa had been preserved from the destructive effects of freezing and thawing by the addition of glycerol. Polge showed that live chicks could be born to hens inseminated with frozen and rethawed sperm preserved in glycerol, and the technique he had used with fowl was transferred to cattle. This was not without difficulty, for bull spermatozoa were, unlike those of fowl, subject to shock on freezing, only overcome by slow cooling. A test with five cows inseminated with frozen and rethawed semen showed that one pregnancy resulted. More extended tests were then made by Rowson at the Cambridge centre in 1952 to achieve a 69% pregnancy rate out of 128 cows using semen that had been stored at $-79\,°C$. More tests followed to ascertain whether pregnancies would ensue with sperm frozen for greater and greater times. They did. The importance of this in breeding programmes was immense. A bull that serves heifers can expect its progeny to complete records of milk production three years later. These can be evaluated and the bull, if he proves superior, used to sire more offspring from a bank of frozen sperm. Subsequently, sperm was stored in liquid nitrogen rather than in carbon dioxide and the cooling process was controlled automatically. Semen was stored in 'straws' to ease insemination. Frozen semen is now in use world-wide as a result of Polge's and Smith's discovery. It was estimated in 1986 that over 20% of all cattle in the world were bred artificially with consequent saving in feed and other resources. The technique also provides some control of disease and in many instances it is associated with constructive breeding programmes. Its importance cannot be stressed too much. It allows the movement of semen around the world; it makes the organisation of artificial insemination programmes for cattle much easier; and although it is not a prerequisite for a progeny testing programme it allows some flexibility over time in the testing and use of bulls.

Similar work on artificial insemination, including cryopreservation of sperm, took place with both pigs and sheep in the 1950s, associated with genetic improvement programmes. This was never on the scale of the cattle programmes. There was one species exception, the turkey. All turkeys are now bred using artificial insemination because the male, as a result of selection for increased growth in size, is physically disadvantaged for mating naturally.

Embryo transfer

It was realised that if, after multiple ovulation, embryos could be transferred between cows (multiple ovulation embryo transfer, M.O.E.T.) a number of advantages would accrue. Embryos of beef calves could be implanted in genetically inferior dairy cows; genetically superior dairy calves could be reared in surrogate mothers of lesser merit, and either of dairy or of beef type; valuable genetic material could, in general, be exported or imported even if there were veterinary health restrictions on the movement of live animals; and valuable genetic material could be multiplied and concentrated, giving powers of selection on the female side almost equal to that on the male side, thus speeding up the selection process. Embryo transfer had been tried before: in 1890 W. Heape had indeed accomplished it. In 1949 the Agricultural Research Council established the Unit of Animal Reproduction under the direction of John Hammond specifically to investigate methods of embryo transfer. This was highly successful in that by 1969 reliable procedures for embryo transfer were first described by L.E.A. Rowson, R.M. Moor and R.A.S. Lawson. The task was difficult. In cattle (unlike in man and some other animals) experimental fertilisation could not take place outside the body; the spermatozoa had to undergo a maturation within the reproductive tract of the female, a process called capacitation. During the early studies the achievement of transfer could only be by the use of embryos, not through the recovery of eggs and their fertilisation *in vitro*. Now problems of *in vitro* fertilisation have been in part overcome. The work has entailed the synchronisation of oestrus of donor and recipient animals; this was accomplished by inducing oestrus by the same means in each, using prostagens or prostaglandin. The donor animal was previously treated with pregnant mares' serum (containing gonadotrophin) to induce superovulation. The gonadotrophic action of pregnant mares' serum had been discovered by H.H. Cole in 1930; although on average it produced eight superovulations it was very variable in its effect. Fertilisation followed, and then there was the problem of recovering the fertilised eggs or embryos. This was accomplished by washing out the fallopian tubes. At first this was done under complete anaesthesia, and later using blocking techniques and laparotomy. A major advance was made when T. Sugie in Japan in 1950 made the first non-surgical transfer. His technique and methods were not very successful. R. Newcomb at Cambridge and J.M. Sreenan in Eire, however, developed methods in which catheters were introduced through the cervix both to wash out the embryos from the donor and to insert them into the recipient. Unlike surgical transfer, when non-surgical transfer is employed the egg must be at the morula or blastocyst stage before it can implant in the uterine wall. This necessitates developing the embryo to this stage, which can be achieved by transferring it at the four- to eight-cell stage to the oviduct of a sheep or rabbit. Trans-cervical recovery of embryos and their insertion into donors using Sreenan's technique are now the methods of choice.

The freezing of embryos

Studies by I. Wilmut in 1972 with early embryos from mice showed that embryos from the one-cell stage to the early blastocyst would survive freezing if they were suspended in media containing dimethyl sulphoxide and were cooled and warmed relatively slowly. In collaboration with Rowson, Wilmut then carried out a series of experiments with cows in which blastocysts that had been frozen slowly, held for six days at the temperature of liquid nitrogen, and then warmed slowly and cultured were inserted into 11 recipient cows. One calf was born in 1973: the famous calf 'Frosty'. Further work followed to obtain reliable methods of freezing and thawing cattle embryos to give acceptable pregnancy rates in recipients; they included the transfer of embryos to straws to aid storage and transport and the use by

Sreenan of the normal insemination gun for the routine transfer of embryos to recipients. The methods of freezing embryos were also adapted to the sheep by J.M. Willadsen and others, and studies were made too with the pig. Towards the end of our period the techniques had developed so far as to allow the splitting of embryos to produce several identical sibling animals.

As a result of this work a number of commercial companies commenced multiple ovulation and embryo transfer as a service to progressive breeders. The reputation of the Cambridge group was extremely high, and it was right that Christopher Polge, who directed the Unit and its successor, Animal Biotechnology, Cambridge, Limited, should have been awarded the Embryo Transfer Pioneer Award. It was appropriate, too, since reproductive studies of animals were started at Cambridge by F.H.A. Marshall, who had studied under Cossar Ewart of telegony fame at Edinburgh, and who influenced John Hammond, who, in turn, opened the way for Polge and his colleagues.

The genetic improvement of livestock

The background and the concepts

The application of genetic thought to the breeding of livestock is closely bound to studies of evolutionary biology and particularly to the work of Sewall Wright, and to a somewhat lesser extent that of R.A. Fisher. Certainly Jay L. Lush of Iowa, who was the forerunner of genetic studies of livestock in the 1930s, was heavily influenced by Sewall Wright, particularly with the development of numerical and statistical methods for the analysis of quantitative traits. In his turn Lush influenced a whole generation of geneticists throughout the world.

Animal breeding is concerned with the choice of parents for the next generation, a choice made with certain objectives in view. This would be simple if the objectives were as simple as those used by Gregor Mendel when he dealt with one gene locus at a time, in crossing his peas. However, most of the traits of interest to animal breeders – milk production, growth rate, body composition, etc. – are controlled by a large number of genes, some with larger effects than others. In any population these genes occur with different frequencies. Quantitative genetics is concerned with the careful dissection of this variation in populations, so as to select as parents for the next generation those that are superior to animals in the population as a whole, to meet the objectives of the programme. Comparison with the development of plant breeding (Chapter 7) is interesting.

Almost all the traits of interest to the animal breeder are influenced by environmental factors as well as by the complex genetic constitution of the animal. Commencing with the genetic component, expressing this as a variance and following the work of Sewall Wright and Lush, a first element can be expressed as being due to the additive effects of genes; that is the amount by which the average of the progeny would differ from the average of the whole population. A second relates to dominance, that is, when the character expressed in the heterozygote is not midway between the expression of the two parental genes. A third, termed the epistatic effect, caters for the interactions between genes. In general, what is manipulated in breeding programmes is the first element, the additive effects of genes; the dominance and epistatic elements are utilised in inbreeding and reciprocal recurrent (i.e. cross testing) programmes, to varying extents, by breeders of poultry for egg production. A consideration of the first element provides the useful concept of heritability (h^2), that is the proportion of the total variation in a population which is due to the genetic element. Its value lies in the fact that it provides a prediction of the genetic gain in performance of stock in a generation when multiplied by the selection pressure exerted. This can then be converted to an annual rate of gain by taking into account the generation intervals. These intervals can overlap as well as differ in length between the progeny of males and females. The reduction in generation intervals was an important feature of the use of performance rather than progeny testing in several species, including the pig (Lush, 1945; Dickerson and Hazel, 1944) and poultry (Lerner, 1950) for which the use of part record selection permitted further reduction in generation interval. The methods of measuring heritability of productive traits were first worked out by Lush in the 1930s. In addition the basic problems of how to predict breeding value from assembled records of stock were understood. Briefly, these group into those based on one or more records of the individual animal, those based on records of ancestors, those based on members of the same family, and those based on progeny.

The American contribution

The immense contributions made to the subject of quantitative inheritance by Lush and his pupils has been mentioned already. Since they were the background upon which British geneticists and animal breeders were to build, some account of them is necessary. Lush, who also has the distinction of probably being the only agricultural scientist to have been born in a log cabin, showed that the performance of an individual animal was a good predictor of breeding value, more so in some instances than a progeny test with limited numbers of offspring, provided that it was selected to correlate well with genotype. With L.N. Hazel at Iowa State College, Lush then expanded the approach to assess the efficiency of different methods of selection. These approaches laid the foundation for performance testing in the United Kingdom. Hazel went on to examine the problem of simultaneous improvement of several traits, to develop ideas about and methods of measurement of genetic correlation. These led to concepts of indices to optimise the ways in which multi-trait improvement of a population could be achieved. C.R. Henderson of Cornell University, who had taken his Ph.D. degree with Lush, developed robust statistical methods to assess genetic parameters as well as approaching solutions to problems of interpretation of progeny tests. At a later date he incorporated his 'Best Linear Unbiased Predictor' (BLUP) in the analysis of progeny test data. This methodology succeeded the previously much used contemporary comparison method devised by Alan Robertson at Edinburgh.

As well as this theoretical and computational work, the American geneticists embarked on experimental studies. Many of these were no doubt influenced by the success of the hybrid maize programmes of the 1930s. G.E. Dickerson developed 38 lines of in-bred pigs; although some showed improvement, most exhibited the common phenomenon of inbreeding depression, with a reduction in litter size and viability which was not reversed by selection. Work also took place in cross-breeding animals to exploit heterosis by breed hybrids and the larger pool of variability that could be drawn upon when different breeds were crossed. This was very successful, although it might be thought only to give scientific respectability to the facts already known to practical breeders.

Post-war developments

As the new methods were being assimilated there was an immense effort throughout the world to try to measure the heritability of production characters in all classes of stock. In the pig, for example, J.W.B. King of the Animal Breeding Research Organisation in Edinburgh measured the heritabilities and genetic correlations in carcases of bacon pigs of both sexes to find high values (about 0.7) for heritability for the depth of back fat at different sites and low values (about 0.1) for litter size. Similar results were obtained elsewhere with, in living animals, very low values for many traits associated with reproduction. In milking cows the heritability of milk production (calculated by Robertson and Rendel) was about 0.25, for liveweight gain in beef cattle about 0.4 and in sheep about 0.15. It will be appreciated that heritability depends upon the variability of the environment as much as it does on the genetic makeup of the population. Under highly standardised managerial and nutritional circumstances the environment is stable and induces little variation, and so the estimate of heritability will be high. Conversely, a variable environment will depress the genetic component. This was of course well realised. So was the implication that livestock breeding had to make the assumption in the long term that the environment did not in fact change from that in which the heritability had been estimated. With these caveats it was possible to provide estimates of heritability and thus of change in commercial qualities in constructive breeding programmes.

Immediately after the Second World War the Agricultural Research Council established an organisation in Edinburgh to work on animal genetics. The Animal Breeding and Genetics Research Organisation had as its first Director Professor R.G. White but included in a rather loose fashion the University Department of Genetics under Professor C.H. Waddington, who succeeded Professor F.E. Crew. On White's retirement the two elements were separated and Professor Hugh Donald became Director of the Animal Breeding Research Organisation with Waddington as Director of the A.R.C. Unit of Animal Genetics and Head of the University Department. The organisations continued to work together very closely. Hugh Donald contributed greatly, not only in his own work on the use of twins for sorting out interactions between genetics and environment and in establishing the value of heterosis in cross-bred animals, but also by his gift for attracting able staff and giving them their head in experimental programmes. These programmes were by their nature always costly, but once convinced he backed his staff all the way and fought for funding in government circles. But perhaps most of all he contributed by his gift for Socratic questioning (apparently at first minatory but in reality very kindly), which stimulated all those who came in contact with him. Waddington had directed operational research in Coastal Command during the war; one of his staff was a chemist, Alan Robertson, whom he invited to change course and become a geneticist at Edinburgh. This was an inspired choice. Robertson visited the United States and worked with both Sewall Wright and Lush. On his return he commenced work with J.M. Rendel on theoretical methods applied to the improvement of milk production in dairy cows. The result was the progeny test, based on contemporary comparisons, which was introduced in 1954. The difficulties that Robertson overcame were that the herds involved were small, the average level of milk production varied considerably between herds and the use of artificial insemination meant that a sire could have daughters in many herds. The basic assumption made in arriving at the contemporary comparison was that the records of a bull's progeny could be expressed as a deviation from the mean of herd mates got by other sires and that the latter were a random sample of bulls other than those being progeny tested. The contemporary comparison test was employed first using the artificial insemination data from Cambridge and then, when the Milk Marketing Board took responsibility for artificial insemination, through the Board's studs. The method was used world-wide.

Robertson predicted at the beginning of the work on contemporary comparisons that the rate of improvement of milk yield would lie between 1.5 and 2.0% per annum in the large population covered by the Milk Marketing Board's artificial insemination scheme. After about 15 years this estimate was not achieved. In fact it was about 0.3%. The reason appeared to lie in the assumption above that the herd mates of the progeny of the bull on test were from a random sample of bulls in the population. More and more, in fact, they came to represent the daughters of proven bulls, thus reducing the value of the comparison. Equally (if not more) important was lack of selection pressure on production traits.

Improvement of pigs

Danish, and indeed most continental European, breeders of pigs used the progeny test as a method of selecting boars to improve their stock. Initially this was done on a relatively small scale in the United Kingdom, but on the advice of J.W.B. King, of the Animal Breeding Research Organisation, the Pig Industry Development Authority and later the Meat and Livestock Commission undertook performance testing, with a pilot scheme being carried out at the University of Newcastle by W. Smith and M. Bichard. The reason for moving to performance testing was that efficiency of food conversion – which has a high heritability – could be measured precisely on the individual animal, while carcase composition can be estimated using ultrasonics to measure thickness of the back fat. The scheme called for submission to central testing stations of litter groups of two boars, one castrated male and one female. Growth rate, fat thickness and food intake were measured on all pigs, and the castrate and female were slaughtered so that additional measurements could be made on the carcases. From the measurements on the boars and those on their full sibs a selection index could be calculated and the boars returned to their owners. These owners were the proprietors of the elite herds of the country and their sales ensured that superior animals flowed to the commercial herds of the country. To provide a measure of progress, two herds – control herds – were set up in which pains were taken to avoid genetic improvement, and which from time to time submitted litters to the testing stations as controls. But changes in the industry undermined this system (Bichard, 1982). The emergence of large companies engaged in the production of hybrid pigs in minimal disease herds (Chapter 11) meant that testing was confined 'in house' to avoid disease problems. But such was the size of these enterprises that they were able to continue similar selection procedures based on the same genetic theory and providing, as a product, superior stock. The success of the procedures is given by the data on the lean content of the carcase at bacon weight in 1974 and 1985, respectively, which were collected by A.J. Kempster and G. Harrington of the Meat and Livestock Commission. The lean content increased over that period of 9 years from 47.6% to 52.1%. The trend continues: in 1989 the lean content of the carcase was estimated to be 56.7%. Current emphasis in pig improvement remains on maximising lean content and on the food requirement to produce it. There is admittedly a complicating factor in that almost half the pigs slaughtered in the United Kingdom are now (1991) intact males, which are naturally leaner than castrates, compared with only 2% in 1975; not all leanness is due to the success of the performance testing scheme.

Although not strictly relevant to the genetic improvement of pigs, a development in the early 1970s is of interest. Messrs. Walls developed a market for a 'heavy hog', a pig to be slaughtered at 118 kg rather than at bacon weight of 90 kg. The idea was that the surplus fat on such a pig, for which there was a market, could be trimmed off and the underlying muscle could be used for a variety of manufacturing processes. The venture was taken up by a large number of farmers because of the attractive price and the lack of need to pay attention to grading standards. However, cheap pig fat from the United States and the high prices paid for the carcases led Messrs. Walls to withdraw. Had the heavy hogs shown lean contents similar to those of modern pigs they might have survived as an alternative outlet for pig meat.

Improvement of sheep

Mention has already been made of efforts to produce new breeds of sheep and changes in cross-breeding policies for producing ewes as dams for fat lamb production. A further development has been of group breeding schemes, to provide flocks of over 1500 ewes from which about 150 individuals are selected as elite mothers and used as the nucleus for the production of superior rams. This scheme was largely developed by Charles Smith of the Animal Breeding Research Organisation as part of the theoretical work on breeding plans that organisation has conducted. National estimates of the body composition of British sheep over a ten-year period ending in 1984 suggest only a small increase in lean content, despite the effects of a good national carcase classification scheme.

Improvement of beef cattle

About half the genes in the beef produced in the United Kingdom originate from dairy cattle, mostly black and whites. Some progeny testing was undertaken by the Milk Marketing Board to test the breeding value of Hereford bulls crossed onto black and white cows, but most of the work to improve beef production has come from performance testing (by the Meat and Livestock Commission) based on live characteristics. Progress has been slow: in a ten-year period ending in 1984, which excluded culled cows, there was a negligible change in the lean content of the carcases.

Theoretical and actual rates of genetic change

At the outset of breeding schemes estimates are made of the rate of genetic change to be achieved by different breeding plans. Many years later these can be compared with accomplishment. An example of this approach was given when the contemporary comparison method for assessing dairy cattle sires was shown to become less effective over the period of examination.

In the poultry industry, selection in broilers was at first mainly for body weight gain and conformation. Increasingly, now, feed conversion is important. Charles Smith examined the effectiveness of selection for weight gain where the expectation had been an increase of 3.2% per annum whereas the accomplishment was considerably greater at 4.1% per annum measured over a year period. There are few other traits or multiple traits which have achieved such good results. Indeed, most long-term studies do not result in the theoretical outcome, particularly when indices are used as selection criteria. The success of selection in the broiler industry may reflect the very simple goals set, their immediate identification with profitability and the very large numbers in the populations undergoing selection.

Major genes in animal breeding

The assumption, which must be true, that complex production traits such as growth, lactation and reproduction involve many genes does not preclude the presence of genes producing major effects. Several of these were discovered during the 50-year period considered in this volume. One which was of considerable interest was the Booroola gene, which imparts large litter size. The study that led to its discovery was made by two farmers, the Sears brothers of Booroola, New South Wales, on their flock of Merino sheep in the 1960s. The brothers maintained a 'multiple birth flock' into which they drafted ewes that had given birth to more than one lamb. Eventually it became apparent that there was a single major gene involved in the improvement in the number of lambs born per ewe. A similar gene was found in the Thoka, an Icelandic sheep, and it appears that a major gene for litter size is also present in the Cambridge, which was described earlier in this chapter.

Questions arose about the desirability of producing high numbers of lambs by increasing litter size under normal seasonal lambing or by producing more than one litter per year with few ewes giving birth to more than two lambs. Both approaches were employed; it appears from observation, nutritional studies and dissections made by J.J. Robinson at the Rowett Institute that physical stress on the ewe carrying four foetuses towards the end of pregnancy is acute, that even with the highest-quality grass available the ewe cannot consume sufficient to maintain her own body weight, and that mortality is high. Mortality of lambs born in litters of four to Cambridge ewes was found to be 47% and for quintuplets 59%. The annual production of 3.7 lambs in a scheme for more frequent breeding does not necessitate large litters and hence does not result in high mortality. In this respect, during the fifty-year period methods for saving lambs were devised based on the syringe administration of colostrum, boxes thermostatically heated to prevent hypothermia, and a number of vaccines to prevent neonatal disease.

When pigs are anaesthetised with the agent halothane, a number of them develop very high temperatures and die. This was traced to a single major gene, the halothane gene, which was found to be widespread in its occurrence in British pigs. The importance of the gene lies in its three effects. Firstly, it increases the yield of lean meat in the carcase; secondly, if the pig is subjected to stress at mating or otherwise it may die suddenly and naturally; thirdly, where the pig is subjected to stress at slaughter it produces a pale exudating muscle, which is unacceptable to consumers. Attempts were made to capitalise on the lean tissue growth rate effects of the gene in the hope that stress-free methods of slaughter would avoid the deleterious effects.

A further major gene is the dwarf gene in poultry; it is sex-linked and used (to some extent) in broiler mother stocks to reduce maternal food costs. Yet another major gene effect, which is very complex in view of its different penetration in different genotypes, is the double muscling or muscular hypertrophy gene. This is commonplace in many continental breeds of cattle, notably the Maine–Anjou and the Belgian Blue. The gene results on average in an increase in the amount of muscle in the carcase by about 15% and a reduction in fat content. The size and shape of the calf at birth, however, gives rise to calving difficulties, particularly where the calf is homozygous for the gene. The difficulties are less where the calf is heterozygous for the gene. In Britain the only major problem from the introduction of continental cattle has arisen from the Belgian Blue, where Caesarean section has to be used frequently, and consequently there has been comparatively little interest in the double-muscled condition. A similar double muscling is found in the Pietrain pig.

Most of the major genes that have been discovered have both advantageous and deleterious effects. Only one was discovered to have none of the latter, the blood group B gene in laying hens, which confers a resistance to Marek's disease.

A bold step for mankind

If we examine the effects of advances in genetics and animal breeding on the development of farming in Britain over the period, we see different changes in different classes of stock. In cattle we see the beef cattle developing within the industry as a series of hybrids, increasingly derived from dairy stock which form the suckler-cow herd for crossing with beef breeds for specialised systems and frequently with continental breeds to produce the final butcher's animal. In the dairy herd we see a gradual replacement of traditional breeds by the Friesian or Holstein and sophisticated methods of selection leading to ever higher production and the very best genetic material being available to all herds by reason of artificial insemination.

In sheep there has long been a series of hybrids derived from the hill sheep and ending with the terminal specialised sire for fat lamb production for the table. What is new is the development of breeding systems, group breeding schemes, and so on, which aim to improve the productivity of the animals, and the development of new breeds for specialised purposes. There is a long way to go in the sheep world. In pigs and poultry we see the almost universal use of commercial hybrid stock; genetic control remains in the hands of a very few companies and the selection procedures are becoming increasingly more sophisticated. Within these trends there is increasing specialisation of breeding stock: broilers versus layers, sire versus dam lines, sows for outdoor/field systems and sows for indoor systems, and the beef breeds and specialised crosses already referred to. The result has been improvement in quality and price in the pig and poultry area and certainly a maintenance of competitiveness in the, as yet, more expensive meat and lamb products.

Selected references and further reading

Bichard, M. (1982). Current developments in pig breeding. *Outlook on Agriculture* **11**, 159–64.

Blaxter, K.L., Kay, R.N.B., Sharman, G.M.A., Cunningham, J.M.M., Eadie, J. and Hamilton, W.J. (1988). *Farming the Red Deer: Final Report of an Investigation by the Rowett Research Institute and the Hill Farming Research Organisation.* Her Majesty's Stationery Office, Edinburgh.

Dickerson, G.E. and Hazel, L.N. (1944). Effectiveness of selection on progeny performance as a supplement to earlier culling in livestock. *Journal of Agricultural Research, Cambridge* **69**, 459–76.

Donald, H.P. and Read, J.L. (1967). The performance of Finnish Landrace Sheep in Britain. *Animal Production* **9**, 471–6.

Falconer, D.S. (1960). *Introduction to Quantitative Genetics.* Oliver and Boyd, Edinburgh.

Fowler, V.R., Bichard, M. and Pease, A. (1976). Objectives in pig breeding. *Animal Production* **23**, 365–87.

Hammond, J., Edwards, J., Rowson, L.E.A. and Walton, A. (1947). *The Artificial Insemination of Cattle.* W. Heffer and Sons Ltd., Cambridge.

Heap, R.B. (1976). A pregnancy test in cows from progesterone in milk. *Journal of the Royal Agricultural Society of England* **137**, 67–76.

Hill, W.G. (1987). *Increasing the Rate of Genetic Gain in Animal Production.* A.S. Nivison Memorial Address, University of New England; Armidale, New South Wales, Australia.

Hill, W.G. (1990). Alan Robertson. *Biographical Memoirs of Fellows of the Royal Society* **36**, 463–88.

Hunter, R.H.F. (1980). *Physiology and Technology of Reproduction in Female Domestic Animals.* Academic Press, London.

Jewell, P. (1983). Rare breeds of domestic livestock as a gene bank. *Proceedings of the Royal Institute of Great Britain* **55**, 209–27.

King, J.W.B. (1981). Animal breeding in Britain 1931-1981. In *The Agricultural Research Council, 1931-1981*, ed. G.W. Cooke, pp. 277–88. Agricultural Research Council, London.

Lerner, I.M. (1950). *Population Genetics and Animal Improvement.* Cambridge University Press.

Lerner, I.M. and Donald, H.P. (1966). *Modern Developments in Animal Breeding.* Academic Press, London.

Longrigg, W. (1976). The place of imported European breeds in British livestock production. *Journal of the Royal Agricultural Society of England* **137**, 42–50.

Lush, J.L. (1945). *Animal Breeding Plans,* 3rd Edition. Iowa State College Press, Ames, Iowa.

Miller, W.C. (1938). Observations on the employment of artificial insemination as applied to British livestock farming. *Veterinary Record* **50**, 25–33.

Newcomb, I. (1979). Surgical and non-surgical transfer of bovine embryos. *Veterinary Record* **105**, 432–4.

O'Connor, L.K. and Willis, M.B. (1967). The effect of artificial insemination on the breed structure of British Friesian cattle. *Animal Production* **9**, 287–93.

Owen, J.B. and Ap Dewi, I. (1988). The Cambridge Sheep – its exploitation for increased efficiency in lamb production. *Journal of Agricultural Science in Finland* **60**, 585–90.

Parkes, A.S. (1985). *Off-beat Biologist.* The Galton Foundation, Cambridge.

Polge, C. and Rowson, L.E.A. (1952). Results with bull semen stored at -70 °C. *Veterinary Record* **64**, 851.

Polge, C., Smith, A.U. and Parkes, A.S. (1949). Revival of spermatozoa after vitrification and dehydration at low temperatures. *Nature, London* **164**, 666.

Provine, W.B. (1986). *Sewall Wright and Evolutionary Biology.* The University of Chicago Press, Chicago, U.S.A.

Robertson, A. and Rendel, J.M. (1950). The use of progeny testing with artificial selection in dairy cattle. *Journal of Genetics* **50**, 21–31.

Robertson, A. and Rendel, J.M. (1954). The performance of heifers got by artificial insemination. *Journal of Agricultural Science, Cambridge* **44**, 184–92.

Robinson, J.J., McDonald, I., Fraser, C. and Crofts, R.M.J. (1977). Studies on reproduction in prolific ewes; I. Growth of the products of conception. *Journal of Agricultural Science, Cambridge* **88**, 539–52.

Ruhane, J. (1988). Review of the use of embryo transfer in the genetic improvement of dairy cattle. *Animal Breeding Abstracts* **56**, 437–46.

Smith, C. (1984). Rates of genetic change in farm livestock. *Research and Development in Agriculture* **1**, 79–85.

Sreenan, J.M. (1978). Non-surgical egg recovery and transfer in the cow. *Veterinary Record* **102**, 58–60.

Willadsen, J.M. (1979). A method for culture of micromanipulated sheep embryos and its use to produce monozygous twins. *Nature, London* **277**, 298.

Wilmut, I. and Rowson, L.E.A. (1973). Experiments on the low temperature preservation of cow embryos. *Veterinary Record* **92**, 686–90.

Wooliams, J.A. and Wilmut, I. (1989). Embryo manipulation in cattle breeding and production. *Animal Production* **48**, 3–30.

11 Animal health and disease

Background

The provision of fodder in early pastoral agriculture depended on the movement of flocks and herds to fresh pasture; at the same time this movement reduced the risks of ingestion of disease-producing organisms and (to a lesser extent) the spread of contagious diseases. The history of animal production since then demonstrates increasing pressure on pasture, as the numbers of people and animals grew and were crammed into the same area of country, with the cycle of use becoming ever more frequent. When migratory, pastoral agriculture gave place to a sedentary agriculture with enclosed fields the same process continued, as stocking rates increased to provide greater productivity; the culmination was the creation of crowded feed lots and intensive housing. With these developments the danger of disease increased; without veterinary help such changes could not have taken place at all. In parallel with increased veterinary understanding, however, have gone the findings of practical people, initiating subtle changes in management practices to avoid disease without a detailed understanding of its nature. Such practices as the frequent strawing-down of yards for the comfort of the animals and the aesthetic satisfaction of their owner also helped to control the ingestion of parasites of the alimentary tract, without a full knowledge of the identity and life history of the parasites. Similarly, rinderpest and bovine pleuropneumonia, periodic serious plagues of cattle in the United Kingdom in the nineteenth century, were both controlled by a veterinary-inspired slaughter and quarantine policy in advance of a full understanding of their aetiology; the former disease was last recorded in the U.K. in 1877 and the latter in 1898. Among the diseases of modern animal husbandry are several that are routinely controlled by simple husbandry measures, sometimes arrived at empirically and sometimes as a result of successful detailed veterinary study. Veterinarians still see these diseases but generally as isolated occurrences, mostly because of some breakdown in the management system.

The epizootic diseases

Foot and mouth disease

There still remain major epizootic diseases that are entirely dependent on veterinary science for their control. The most feared disease at the beginning of the modern agricultural revolution was undoubtedly foot and mouth disease, which attacks all cloven-hooved species (cattle, sheep, pigs and goats). A highly contagious epizootic disease, it is present in most countries of the world, except those that have taken stringent measures to eliminate it. The condition is recognised by the development of blisters on the tongue and lips and on other parts of the body where the skin is thin, including that between the toes of the ungulate hoof, hence its name; a number of other infections are only distinguished from the disease with difficulty.

Infection is followed by a rapid febrile illness, in the course of which the virus (the particles of which are very small and resistant to destruction) is released in very large amounts in the bodily secretions, from pustules and in the copious salivation. Spread is rapid and over comparatively large distances. The mortality caused by mild strains may be as low as 5%, that of severe strains as high as 50%. In surviving milk animals there is a sharp drop in milk yield and in fattening animals a severe loss of condition. It therefore became a policy in all advanced agricultural countries to try to contain the disease by a draconian slaughter policy. On notification, all affected cloven-hooved animals were slaughtered along with all similar contact animals (defined as animals on the same farm or in a wider infested area). The movement off the farm of all animals, persons or things was prohibited and the entry of persons onto the farm severely restricted. It is almost impossible to exaggerate the fear engendered by the disease, resulting from its sudden appearance and reappearance; the horror to the farmer and his family and the desolation that ensued when all the stock on the farm had been slaughtered; the knife-edge of uncertainty for farmers on contiguous farms, themselves free from the disease; the heavy responsibility of the veterinarians called upon for diagnosis and supervision of slaughter; the funeral pyres and mass graves to which whole herds of high-quality cattle, sheep or pigs were consigned; all of these and the sense of isolation and anxiety experienced by those on infected farms made the announcement of the re-appearance of foot and mouth disease in Britain strike terror into the heart of the farming community.

The logistics of research on this disease were difficult because of its highly infectious nature. The first effort to get round the problem was made in Britain in 1912 when a research facility was set up in India, where the disease was endemic. Problems of staffing and provisioning at a distance caused this effort to fail after a very few months. The next attempt was to set up an isolation centre on a obsolete warship and an associated lighter, at Harwich in 1920; this failed also. The very severe epizootic from 1922 to 1924 led to the courageous decision to set up the Pirbright Experiment Station in Surrey, in strict isolation, in the former Ministry of Agriculture cattle testing station. The history of the disease in Britain was then one of increasing understanding, improving techniques, and eventual control. First came the development of a method of quantitative determination of the titre of the virus by Sir William

Henderson, using inoculation of serial dilutions on the tongues of susceptible animals and counting the developing lesions. Then came improvements in serological identification and the demonstration of the susceptibility of unweaned mice, which allowed more extensive and more detailed quantitative work. The subsequent development of tissue culture methods allowed a whole new step forward in accuracy, ease of handling and identification of the virus. The demonstration in 1960 of the value of baby hamster kidney tissue culture for the multiplication of the virus opened the way for the commercial production of vaccines against the disease, in which Pirbright took a leading part. Another element that contributed to the control of foot and mouth disease in Britain was the discovery that alkaline solutions, such as sodium hydroxide and sodium carbonate, were more effective disinfectants of the virus than the usual phenolic materials used on the farm. This was important because of the capacity of the virus to persist on bedding, walls, trucks, etc. Next, the discovery that the virus could survive in chilled meat and offal pinpointed a recurrent source of possible reinfection. Meat from South America, where the disease was endemic, was a particular threat. When it was understood that the actual edible portions of meat were cooked and effectively sterilised, it was realised that bones, which might be discarded, or fed to dogs in a raw state and carried round the farm yard, or fed to pigs in swill, were the most likely sources of reinfection. Accordingly the prohibition of meat imports from potentially infected areas was modified to allow only the importation of boned meat products. Good circumstantial evidence suggested that the disease could be carried by the wind over considerable distances, but this was probably more important for within-country spread during an epidemic. Vaccination of animals as a control measure was never allowed in Britain, but the success of vaccination programmes in other countries reduced the amount of virus threatening reinvasion of the British Isles and contributed to the success of the overall control measures.

Most importantly, the speed and precision of modern laboratory techniques for the identification of the virus and its differentiation from other diseases, that may simulate it allow a slaughter policy to be put in place when the epizootic is at an early stage and still limited in scale, with correspondingly less distress in the farming community. The disease is no longer the mysterious threat that it was, but such is its nature that we can never rule out the possibility that it will reappear.

Tuberculosis

Tuberculosis of farm animals was another disease of great importance between the two World Wars. For many years its true importance was overshadowed by tuberculosis of human beings, which in the early years of this century was the chief cause of death in Britain, western Europe and the

United States of America, with an annual death rate of nearly 200 per 100 000 persons. The human form was a respiratory disease spread in overcrowded and ill-ventilated conditions, which sometimes spread to other parts of the body by reinfection; it gradually responded to increased knowledge, improved conditions, new drugs and improved management regimes for the sick, so that in Great Britain the annual death rate per hundred thousand had dropped from 190 in 1900 to six in the 1960s.

Meanwhile bovine tuberculosis, which is caused by an antigenically distinct form of *Mycobacterium tuberculosis*, gradually came to be recognised not only as a serious disease of cattle but also as the cause of important clinical conditions in the human. In the U.S.A. in 1918, about 5% of the herds of milking cattle were known to be affected, that is to say, they reacted to the tuberculin test. By 1950 the number of reacting herds had fallen to 0.2%; a similar picture was shown in the U.K. The Gowland Hopkins Committee, reporting in 1934, accepted that at least 40% of cows in the U.K. were, or had been, infected with tuberculosis and that 0.5% of all cows yielded tuberculous milk (Pearce, Pugh and Ritchie, 1965). These statistics concealed a potentially serious situation, for individual infected cows with mammary lesions could contaminate a great deal of milk. Where milk was pooled from one herd or from a district in the absence of pasteurisation, the chances of infection from drinking milk were comparatively high. The ingested bacteria invaded the lymph glands, particularly in children, and could attack any of the organs of the body: bones (an important cause of lameness and spinal deformity), kidneys, liver, and the lining of the brain (to cause meningitis). A more widespread and generally fatal infection of the lungs, mesenteries, etc., was referred to as miliary tuberculosis. Although, except in the case of meningitis and miliary tuberculosis, the death rate from infection with bovine tuberculosis was less than with the human variety, it was frequently damaging, crippling and disfiguring and always to be taken seriously.

The veterinary control of tuberculosis in cattle relies less on spectacular research discoveries than on the systematic development of statutory methods of control by the veterinary officers of agricultural departments. Compulsory slaughter of animals showing overt clinical signs of tuberculosis became possible under the Tuberculosis Orders of 1925 and 1938. Its value was more in the area of public health than in eradication. Eradication came to depend on the testing of cattle by the use of tuberculin, discovered first by Robert Koch in 1890 and then in a purified form by Seibert in 1934. Tuberculin, a protein from the coat of the tubercle bacillus, sets up an inflammatory reaction when injected under the skin of cattle that have contracted tuberculosis prior to the injection (a parallel reaction in humans forms the basis of the Mantoux test). The Gowland Hopkins Committee, reporting in 1934, recommended the development of a scheme of voluntary testing of herds with tuberculin to create attested herds. Owners of herds that had passed three tuberculin tests were allowed to sell tuberculin tested (T.T.) milk and were awarded a government premium of 1d (0.4 new pence) per gallon, which in 1943 became 4d (1.6 new pence) per gallon. There were also bonuses of £1.00 per head for tested disease free cattle in the years to 1950 and by 1960 it was possible to declare the whole country attested, with the number of cattle slaughtered under the Tuberculosis Order down to 7 in 1960 compared with 22 000 in 1935. At the same time there was a parallel development of pasteurisation, which was slow to take off because the technology did not produce milk with consistently good flavour; however, the development of 'flash' pasteurisation, which produced a high-quality product, finally allowed the Ministry to introduce mandatory pasteurisation in the 1970s. These two measures together have reduced infection with bovine tuberculosis in humans to a low level. The discovery of the antibiotic streptomycin by Waksman in 1943–44, and the complementary drugs para-aminosalicylic acid and isonicotinic acid hydrazide in 1946 and 1952, provided other weapons in the fight against tuberculosis of both types in human beings. Testing with tuberculin and vaccination also helped to control the diseases although at the time of writing there is concern that human tuberculosis is showing resurgence in conditions of poverty and overcrowding, and most recently with the emergence of human immune deficiency disease (AIDS).

Tuberculosis also occurs in poultry (avian tuberculosis) and in pigs, which are affected by the avian, bovine and human forms of the disease. One other important control measure, which does not receive as much recognition as it deserves, is the *post mortem* veterinary inspection of carcases intended for human consumption. The identification and disposal of carcases carrying tuberculosis lesions is a very important safeguard for the meat-eating public.

Cattle diseases

Diseases of adult cattle

With control of these important diseases of cattle in sight, the United Kingdom Ministry of Agriculture sought to identify the priorities for the next phase of the control of economically important animal diseases. Four areas were designated in cattle in 1940 as having high priority: mastitis, brucellosis, Johne's disease and poor reproductive performance.

Mastitis, an infection of the mammary gland by Streptococci, Staphylococci or *Escherichia coli*, was in the 1930s a well-recognised condition in lactating mammals in general and in particular a troublesome economic impediment to the production of milk by the cow, particularly the dairy cow. Every increase in the intensification of milk production, larger herds, loose housing, machine milking, the bulking of milk and its statutory sale to the Milk Marketing Board

(1933), was paralleled by an increase in the importance of subclinical mastitis. Often as many as 50% of cows in a herd would suffer from the subclinical form, recognised by an increase in the count of leucocytes in the milk in the regular monitoring and quality control procedures. This was a difficult and sometimes intractable problem, which responded only in part to the efforts of veterinarians and practical farmers. Credit must go to the National Institute for Research in Dairying where Frank Dodd and his colleagues (Thiel and Dodd, 1977, 1979) were at the centre of a national effort which involved research workers in a number of disciplines, workers in the development services of ADAS and the Scottish Colleges, and practical farmers. The important elements in control were: firstly, the development of a systematic antibiotic treatment of the udders of cows during the drying-off period, when the antibiotic was introduced into the teat duct; secondly, the introduction of a twice daily teat dipping to prevent infection and the spread of infection; and thirdly, a study of the design and functioning of the double-action milking machine. This last was important because there had been a tenfold increase in machine milking from the 1930s to the 1980s while the number of cows within the herd had increased perhaps fivefold and production per cow had at least doubled with presumptive evidence that mismanagement of the milking process was an important element in the disease aetiology. Research into the design of the teat cup and liners showed that efficient removal of milk from the teat was not always achieved, and X-ray cinematography of the milking process allowed the development of a design that minimised damage to and reinfection of the teat. Ely and Petersen's (1941) discovery of the neurohormonal control of milk secretion coincided with the adoption of an udder-washing technique which, at one and the same time, cleansed the udder and substituted for the nuzzling stimulus of the calf, facilitating milk let-down and reducing milking trauma. These separate research findings and the integrated programmes of development they generated on the farms of the Agricultural Development and Advisory Service and the Scottish Agricultural Colleges and of private farmers has brought mastitis control to a reasonable level. Constant vigilance remains necessary, however, since in loose housing conditions sawdust bedding is thought to encourage reinfection, and slatted and scraped floors encourage coliform mastitis, which is often acute and sometimes fatal. With this disease scientists have provided answers that would allow full control of the disease. Economic factors prevent its complete achievement. The separate problem of summer mastitis in suckler herds still remains a sporadic and intractable condition.

Brucellosis, caused by the bacterial genus *Brucella*, occurs in at least three species of farm livestock: in goats and sheep as *Brucella melitensis*, in swine as *B. suis*, and in cattle as *B. abortus*. All can be transmitted to humans. It is the last named that has been of particular importance to the U.K. farming industry in the period under review. In fact the disease in goats was first recognised by

Sir James Bruce, who isolated *Brucella melitensis* from a victim of the disease in Malta, and who gave his name to the bacterial genus. In the U.K. and the U.S.A. the disease of cattle has long been recognised as a considerable tax on the animal industry and as a very serious threat to human health. In infected cattle the organism occurs in the milk, in aborted foetuses, placental membranes and fluids, and from the persistent vaginal discharges that occur for several weeks after an aborted birth. The organism is also present in the semen of infected bulls, although there is no evidence that the disease is passed from bull to cow in natural service. Although infection of humans may occur through drinking milk, most transmission is from calving cows; veterinarians and stockmen handling aborting cows at or after parturition, and slaughtermen dealing with infected carcases, are most at risk. Brucellosis in humans in its acute form may start abruptly with a severe fever, which declines to give spontaneous recovery after some months, if not treated. The disease in its chronic form often develops insidiously, with a low-grade fever, chills, weakness and loss of appetite, and generalised aches. It may go unrecognised for some time. The acute form responds to treatment over several weeks with antibiotics. The chronic form, which is more commonly (but not invariably) caused by *B. melitensis*, is less easily treated, and those affected suffer a serious chronic illness with intermittent recurrent symptoms and sometimes neurological and psychiatric complications.

In cattle, attempts were made to control the disease on an individual herd basis in Britain and the United States by the eradication of infected cattle, at the very beginning of our fifty-year period. It often proved a thankless task. Herds that were brucellosis-free, in a situation where the organism was still frequent in the cattle population as a whole, were very vulnerable if reinfected. Under these conditions so-called abortion storms would take place, with a large number of cows aborting in one herd near the normal calving time, with great economic loss. On a country-wide basis, the main thrust of eradication came after the end of the Second World War.

Work at the Institute for Research on Animal Diseases at Compton Manor, Berkshire, from 1949 to 1956, under conditions of strict isolation, showed that two vaccines, S19 (developed in the United States) and 45/20, both of which used attenuated strains of the organism, were effective against the disease. S19 appeared to be more stable and was preferred. Thereafter a three-pronged attack was made on the disease: by vaccination of calves; by the identification of infected cattle using a serum agglutination test; and by the slaughter of proved reactors. Great credit must go to the state veterinary service and the veterinary investigation service for the dedicated work carried out in implementing the testing procedures. The presence of the antibodies to the S19 vaccine required careful discrimination in reading the tests and the use of a dilution technique to ascertain the titre level to avoid false positives. Other practical aids were the development of mechanical methods for the titration of the samples to speed up the process of identifying reactors, by workers in the Scottish Colleges of Agriculture and the Ministry of Agriculture Veterinary Laboratories.

This campaign, which epitomises the obligatory relationship between science and practice in development work, depended for its success on the careful work of a number of clinicians and bacteriologists, the development of effective vaccines, the identification of reactors by serological methods in a logistically demanding programme, and the administration of sensible government legislation for a slaughter policy. The removal of this disease from the national herd has not only been of great economic benefit but has removed a severe health risk from those who work with cattle.

Johne's disease, a wasting disease of cattle caused by *Mycobacterium paratuberculosis*, is neither so widespread nor so dramatic in its manifestations as the foregoing, and similar spectacular advances in its control have not yet been made. The disease tends to recur within a herd; cows may remain infected for several years before the signs of the disease become obvious. Research findings during our revolutionary period opened the way for the eventual removal of the disease from the national herd. Firstly, it was discovered at Compton, in animals kept under strict isolation, that only young animals could become infected (Burns, 1981). Secondly, it was found that the isolation of the organism, previously a matter of difficulty, could be facilitated by the addition to the isolation medium of a growth factor extracted from cultures of *Mycobacterium phlei*, a plant pathogen. Thirdly, the isolation of Johnin, which is produced in the same manner as tuberculin, produces a similar reaction and has been a great help to those seeking to identify and eradicate the disease.

Infertility in cattle is not a single clinical entity but a problem which occurs for a wide variety of reasons. Failure to breed not only represents an economic loss in terms of the time required to produce a given weight of saleable product but also represents an impediment to economic management. A herd of cows, for example, calving at one period of the year is easier to manage than one in which calves are produced irregularly throughout the year and in which medication, levels of feeding, marketing, etc. have to be staggered to meet the different age groups. Some infertility is attributable to reproductive diseases caused by organisms such as *Campylobacter foetus* and can be controlled by testing programmes. However, there is a wide range of clinical conditions that reduce fertility; other factors, such as nutrition and management, interact with them. Investigations continue to assess their relative importance.

Respiratory and diarrhoeal diseases of calves

Two other groups of diseases were particularly important in reducing the productivity of cattle enterprises during our fifty-year period. They were the respiratory diseases and the diarrhoeal diseases of calves. Although great advances have been made in the understanding of these diseases, in the isolation of bacteria and viruses associated with them, with the development of gnotobiotic (disease-free) calves for research purposes and the re-creation of the clinical disease, control is still not simple.

In the respiratory diseases a number of monovalent vaccines have been produced and the importance of poor ventilation as a pre-disposing factor has been agreed. Housing design has been an important research and development activity; various methods of ensuring adequate ventilation have been tested.

Diarrhoea in calves, sometimes caused by strains of *Escherichia coli*, is still an important cause of loss and there is increasing evidence for other diarrhoea-producing pathogens, both viral and protozoal. Although various curative measures became available throughout and particularly towards the end of our period, the overwhelming importance of hygienic management of calves led to further improvements in the design of calf cubicles and housing and attention to the time of housing of calving cows, etc.

Sheep diseases

The second most important animal species in the United Kingdom is the sheep. The industry is based on an inter-dependent stratified system of production from the highlands to the lowlands. Central to the success of the system is the provision of adequate winter nutrition and the control of the main diseases that affect productivity. With a bleak smile farmers say that the sheep is an animal 'whose main purpose in life is to die'. The economic pressures on the sheep industry in recent years have demanded higher productivity deriving from higher stocking rates, higher lambing percentages, faster growth rates, and manipulations such as attempted twice-a-year lambing. These have all depended on concomitant advances in veterinary science.

Clostridial diseases

It was the recognition of the importance of sheep diseases that led a number of Scottish farmers to set up the Animal Diseases Research Association in 1921–22, at first funding work by Gaiger and Dalling at the Edinburgh and Glasgow Veterinary Schools, and then in 1926, with help from the Development Fund, establishing its own premises at Moredun, Edinburgh. From the very beginning a determined attack was made on the clostridial diseases, which were recognised as one of the main problems in sheep husbandry. Clostridia (bacteria of the genus *Clostridium*) had been identified in the later years of the nineteenth century as being associated with a number of important diseases of sheep (Sterne, 1981; Buxton, 1983) although in the absence of detailed work there was an inevitable lack of security in the diagnosis and attribution to specific organisms. The Moredun work, which impinged on our period, clearly demonstrated the range of clinical entities – blackleg, braxy, black disease, enterotoxaemias including lamb dysentery, struck, pulpy kidney – and identified the associated clostridia. Following from this a range of vaccines was produced and these or their derivatives are now used routinely to control the diseases. An interesting aspect of the vaccine work was that although the expertise to produce new vaccines lay with the research worker and the veterinary research laboratories like the Moredun, Pirbright and Compton, as soon as the methodology was systematised and shown to be safe it was taken over by commercial organisations, which are the main vehicles for production today. Vaccination for disease is an important element in sheep husbandry. Vaccination schedules have been worked out and are followed rigorously in well-managed flocks (Mathieson, 1983).

Trematodes and nematodes

Sheep at pasture are very susceptible to the attacks of helminths, both trematodes (flukes) and nematodes (worms). Many workers have contributed to the elucidation of the life histories of these organisms, which sometimes (flukes) involve a number of stages and alternative hosts (Coop and Christie, 1983; Whitelaw, 1983). It would be fair to say that the zoological detail had mostly been determined before our fifty-year period but that the discovery and development of practical nematicides and flukicides falls squarely in it. The nematicides include levamisole, benzimidazoles and ivermectin (Chapter 6). In the case of flukes the first substance used was carbon tetrachloride, which had to be administered with care because of its poisonous nature. Later came diamphenethide, nitroxynil and rafoxanide. The most interesting outcome of the success of these treatments has been the development of management systems and forecasting systems which, in combination with strategic and more economical use of drugs, provide good control of the parasites.

In the case of liver fluke (*Fasciola hepatica*) the ecological work of Ollerenshaw, Ross and others (see Whitelaw, 1983) led to the development of forecasting systems relating rainfall to the presence of infective liver fluke larvae on the pasture, thus allowing variations in the dosing regimes. Another approach (Armour, 1983; Whitelaw, 1983 and others), sought to kill the flukes in the host at an immature stage so that eggs were never deposited and the presence or absence of snails, the alternative host, became irrelevant. This method, used since 1973 by Whitelaw at the Hill Farming Research Organisation's Lephinmore farm in Argyllshire, led to the virtual elimination of the fluke.

The control of the nematode worms in sheep follows two routes: control by dosing with recommended nematicides, and control derived from a knowledge of the natural history and life cycles of the nematodes. Control is sometimes achieved by a mixture of both.

In nematodiriasis, caused by *Nematodirus battus* in lambs, the infection is carried from lamb crop to lamb crop, the worm eggs surviving in the pasture from spring to the following spring, when they are activated by changes in the climate. The disease can always be avoided by ensuring that lambs are grazing on grass that has not carried lambs in the previous year – a counsel of perfection on most farms – or by strategic dosing with anthelmintic drugs following prediction of the disease risk from April to June. To hatch, the worm eggs require a period of chill followed by a mean day–night temperature of more than 10°C. It is thus possible to forecast the date of the spring hatch, and to modify the three-weekly pattern of dosing accordingly.

Another type of worm infection in sheep is parasitic gastroenteritis in lambs, caused by a number of nematode genera but particularly by species of *Ostertagia* and *Trichostrongylus*. The eggs of these worms are passed in the faeces of the ewe in large numbers during the periparturient period, when there is a relaxation of immunity. Effective dosing of the pregnant ewe followed by a further treatment immediately following lambing is the method of control commonly used. Eggs deposited in the first half of the grazing season, April–June, are responsible for the potentially dangerous populations of larvae that accumulate in the second half of the grazing season, July–September (Armour, 1983). Knowledge of the life history of the worms and the availability of effective nematicides has allowed the development of a number of ingenious systems of management, which economise on dosing and produce good results. When grazing is plentiful, dosing ewes immediately after lambing before moving onto fresh pasture reduces the risk to lambs. Extending this further, the ewes may be wormed and strip-grazed, with the lambs being allowed to graze ahead of the ewes. A further refinement is the Forward Creep Grazing System developed by Cooper, Dickson and Black at Newcastle (Black, 1960) and a similar system developed by Spedding (1965). It was during the development of these systems that the importance of totally fresh pasture, to avoid nematodiriasis, was appreciated. The system as finally evolved was effective but expensive. However, it put in place the ideas for the Clean Grazing System devised by Rutter (1975) of the Scottish Agricultural College. This is economical of labour and of drugs but calls for a high degree of management skill. It can only be used in an alternate husbandry situation where a grass ley is renewed at one point in the system. A field that has been devoted to sheep grazing is ploughed for a cereal crop undersown with grass or for direct-sown grass and clover. A cereal crop or a hay crop is taken from this field. In the following year the field is given over to (for example) cattle grazing and in the following year sheep and lambs are introduced, the ewes being wormed as they leave the lambing shed. Under this system the summer stocking rate is determined by the quality of the pasture and the manurial programme; it is surprisingly high (17 ewes with 29 lambs per hectare maintained as a demonstration over a five-year period). Careful attention must be given to area equivalence from year to year so that the stocking rates match; some kind of escape mechanism to provide alternative worm-free pasture in times of drought is useful. This system and adaptations of it owe as much to developments in grassland management as to those of the veterinary parasitologist, but they could not have been contemplated without an effective worm treatment.

Other diseases of sheep

An increased understanding of the many other diseases of sheep has led to their control by sensible husbandry measures, by the development of vaccines and curative drugs and by the use of antibiotics. Many of these diseases are sporadic or localised. For example, louping ill, a virus disease of the central nervous system, is tick-borne and localised to certain hill areas, where it may also attack grouse. Where the disease is endemic, vaccination is necessary and effective. Enzootic abortion is endemic in many low-ground flocks. Work by John Stamp and colleagues (Aitken, 1983) demonstrated that it is caused by the protozoan organism *Chlamydia* and produced a successful vaccination procedure. Several other abortion diseases of sheep are currently under investigation: border disease (a virus), campylobacter abortion, and toxoplasmosis, caused by the protozoan *Toxoplasma gondii*. Toxoplasmosis is of particular interest because fresh outbreaks in hitherto uninfected sheep flocks can lead to early lambing of weakly lambs, abortions, and the production of mummified lambs, depending on the stage of infection of the ewes. Its life history and pathology were worked out in our period (Frenkel, 1971) but understanding of the disease is still expanding. The primary host is the cat, which sheds resistant oocysts into the environment, where they may be picked up by sheep, birds, mice or even human beings. Cats acquire the organism in general by eating mice; in the early stages of infection, young cats excrete large numbers of oocysts. Feed stuff may become contaminated and, when fed to sheep during pregnancy, may be the source of infection. Although infection of cats is sporadic the chance of an epidemic situation arising from an infected cat increases with time and with the number of cats around. It is also important that pregnant women take care in situations where chlamydial abortion or toxoplasmosis are present.

Scrapie and bovine spongiform encephalopathy

One other disease of sheep, a potential threat to human health, is scrapie (Mitchell and Stamp, 1983; Dickinson, Stamp and Renwick, 1974). It must be stressed that whatever the *potential* threat may be there is as yet absolutely no evidence of the transmission of this disease to human beings. It is a disease of sheep caused by a so-called slow virus, because the incubation period is very long, sometimes as long as the lifetime of the animal. The symptoms of the disease are loss of condition, wasting and a nervous irritability, and pruritus, which is manifest in much rubbing and scratching of the body on adjacent fence posts, wire, etc. There is also evidence of a high-stepping gait in the forelegs, which leads to incoordination in all four legs. The animal eventually dies.

The causative organism is unusual, first in terms of the long incubation period, and then in the fact that it is very small, has great resistance to heat and does not provoke an immune response in the host. A great deal of work has been carried out on the disease because of its similarity to certain encephalopathies in humans. An advance has been made in demonstrating its transmission to mice, where it is more easily handled than in sheep. There is evidence of genetic differences in the reaction of the host to infection, in both sheep and mice. Strains of sheep can be produced which are resistant to the disease in so far as they fail to show the signs of the disease during a normal life span. Similarly in mice, where the experimental evidence is clear for strains of mice and scrapie, in some combinations the incubation period is greater than the life span of the mouse. Accordingly there is a general policy of culling infected animals and of attempting to remove the disease from the national flock. Much of our knowledge of the disease is insecure but it does seem that there is maternal transmission of the disease and that it is not through the milk. The evidence that the pasture may be contaminated and the disease picked up from the pasture in the absence of animal hosts is more doubtful. It is clear, however, that goats, which may be affected also, have become diseased after mixing with a flock of scrapie-infected sheep.

Diagnosis in the live animal is difficult but the post-mortem findings are said to be quite clear, with vacuolation of the neurones, particularly in the *medulla pons* and the midbrain, and with a spongiform vacuolation of the interstitial tissue of the brain.

These post-mortem findings have focused attention on the disease, which bears pathological similarities to two human diseases, kuru and Creutzfeldt–Jakob disease. Interest has been heightened with the recent appearance of 'mad cow disease' (bovine spongiform encephalopathy), a condition strikingly similar, indeed probably identical, and thought to have been contracted by calves eating a high-protein feed supplement which derived in part from carcases of sheep, including some with scrapie. The first cases of the disease were confirmed microscopically in November 1986, which comes within our own 50 year period. It is probable that the earliest clinical cases occurred in April 1985. Since most cattle are now known to be infected as calves through the animal protein in their diet, and the disease has commonly an incubation period of about 4–5 years, it is likely that the first infections took place in the early 1980s (Hope, 1994). By January 1994 nearly 117 000 cases had been confirmed. It is now thought that the reason for the sudden onset of the epidemic was a change in the method of rendering the ruminant carcases incorporated in the feed, a small proportion of which carried the scrapie virus from infected sheep. An upsurge in the disease which occurred in 1989 is considered to have been due to the inadvertent inclusion of infected cattle carcases in the processed protein feed. A ban on ruminant protein in ruminant feed was instituted in July 1988.

The progress of the epizootic has been very closely monitored and it appears that while there was an increase in confirmed cases from 1986 to 1993, after that date there was a decline in the number of two-year-old cattle confirmed as diseased. At the time of writing (1994) the number of new cases in that age group is zero. Accordingly, there appears to be the beginning of a decline in the reputed new cases and the hope is that the infection rate will fall to zero. The problem remains of the possibility of mother-to-daughter transmission or of horizontal transmission between cattle in a herd. This is currently the subject of detailed analysis and experimentation, but the negative findings to date suggest that such transmission is likely to be negligible compared with transmission through the feed supply.

With regard to human health, the severe measures instituted for the destruction by high-temperature incineration of clinically diseased cattle has removed a major source of potential infection. Further, the measures taken to exclude nervous tissue of healthy (potentially infected) animals from the human food chain suggest that fears of consequences to humans from the cattle epidemic are probably without foundation. Barlow (1991), who has been responsible for much of the experimental work on the passage of the disease agent from cattle to mice and between mice, is optimistic. He writes, 'The early results of work specifically designed to test the infectivity of a range of tissues from cattle terminally ill with bovine spongiform encephalopathy have revealed no evidence suggestive of extra-neural accumulation of the agent. This suggests that the fear of risk to human health from consumption of dairy products and meat is unfounded.' This is supported by the historical evidence from scrapie. Scrapie has probably been around for many centuries, but no obvious connection has appeared with any human conditions. Moreover in Australia and New Zealand, where scrapie is absent, the incidence of spongiform encephalopathies such as Creutzfeldt–Jakob disease is the same as that in the rest of the world, where it appears as one case in 1–2 million.

The public perception of the disease has not been favourable to the meat industry; the situation can only be resolved by further scientific work of a high standard.

Pig diseases

By the end of our fifty-year period the U.K. pig industry was producing some 15 million slaughter pigs per annum. The average herd size was still small, with about 30 sows, and the total number of pigs per unit was a little more than 300. Throughout the period economic pressure pushed up the numbers in many more production units, particularly finishing units with intensive housing. These units might require to draw their weaner pigs from a wide area with resulting mixing of batches of pigs. Breeding units were different. The production of elite stock was in the hands of pedigree breeders, who submitted samples for central testing. There were obvious health dangers here. Then, as the production of superior stock passed into the hands of a small number of breeding companies producing hybrid pigs of high health status and derived from minimal-disease units, the problem of disease spread was much reduced.

A wide variety of diseases affects the pig. Many of these parallel the diseases of sheep or cattle, e.g. various diarrhoeal and respiratory diseases. Others are distinct and different. Swine fever (hog cholera in the U.S.A.) is a notifiable disease and has been controlled in Britain by a stringent isolation policy and slaughter of all infected and contact animals. It was eradicated from Britain in 1966 but there have been recurrent outbreaks, e.g. near Hull in 1971 on farms where swill was fed. The disease is now absent from the U.K., Ireland, Scandinavian countries and the U.S.A., Canada and Australasia. However, it is present in other European countries and must always pose the threat of reimportation to Britain.

Specific pneumonia-free pigs

Two important diseases were the subject of novel approaches to control. The first, enzootic pneumonia, a highly contagious pulmonary disease caused by *Mycoplasma hypopneumoniae*, occurs endemically in the general pig population where it causes a low-grade pneumonia, with unthriftiness and coughing. In an attempt to eliminate it from the pig population the idea was promoted of removing piglets from the dam at term, by hysterectomy or hysterotomy (Twiehaus and Underdahl, 1970; Taylor, 1979). The piglets so obtained were dropped through an antiseptic curtain into a clean environment, where they were reared out of contact with other pigs. From these piglets a specific pneumonia free (S.P.F.) herd was developed. The process provided freedom, also, from most infectious diseases of the pig. The nucleus herds of large companies producing hybrid pigs are mostly enzootic pneumonia-free and are maintained as closed herds with strict control of the entry of new blood lines, which must themselves be from hysterectomised sows, although in some circumstances semen and piglets from hysterotomised sows may be admitted.

Aujesky's disease

The second major pig disease, Aujesky's disease, is worthy of notice because an attempt has been made to eradicate it in the United Kingdom by using a producer-funded eradication scheme. It is caused by a herpes virus with severe manifestations in younger pigs: diarrhoea, vomiting, trembling, incoordination, spasms and frequently death. In older pigs the manifestations may be less severe but with an extension of the finishing period; in adults the signs are usually transitory but reproductive problems may follow. The disease was made notifiable in the U.K. in 1979 and movement of affected pigs controlled under an order of 1982. The main eradication programme, with producer cooperation and funding as well as help from the M.A.F.F., took place over the summer and winter of 1983–84, when some 470 herds were slaughtered. Thereafter attention was concentrated on the serological identification of affected pigs, with slaughter following where more than 10% of a herd was identified as positive. Where less than 10% of positives were found slaughter only followed where subsequent transmission was demonstrated. The campaign has been successful.

Mycotoxins

A final problem in pigs arises from their being simple-stomached animals. The identification of mycotoxins such as the aflatoxins affecting chickens (Chapter 9) led veterinarians to keep a look-out for fungal contamination of the feed stock which might be responsible for disease problems in pigs (Robb, 1988). Among the early identifications were ochratoxin A (produced by *Aspergillus ochraceus*) and zearalenone (produced by *Fusarium* spp.), which are responsible for kidney degeneration and infertility, respectively, in the pig.

Poultry

Changes in husbandry

Poultry husbandry changed out of all recognition over the period of our survey. After the First World War poultry keeping, which had previously been a small-time activity (often of farmers' wives) in Britain, was encouraged as a useful smallholding activity. Ex-servicemen then and unemployed people during the Depression took up poultry keeping as a source of income. For obvious reasons many of these people were unskilled in the handling of animals, and disease became important. Enterprises grew in scale but were still largely outdoor and limited by the physical labour required in moving arks, etc. The veterinary profession was slow to become involved. After the Second World War larger-scale enterprises began to be developed in Britain and throughout Europe encouraged by development work in the U.S.A. These used deep-litter systems and caged systems, often tiered as 'battery cages', which enabled very large numbers of birds to be kept in controlled environments, with automated feeding, watering and egg collection by fewer workers. Sainsbury (1980) presents data showing how the percentage of the various systems in Britain changed over the period 1960–77. The proportion of units using battery cages rose from 19.3% in 1960–61 to 93.2% in 1976–77, and in parallel the average number of eggs laid annually per bird, for all systems (free range, battery cages and deep litter) rose from 185 in 1960–61 to 243 in 1976–77. Following their success in breeding better egg producers, poultry geneticists began to take an interest in breeding meat-producing birds, which grew faster and utilised their food better. They could also be managed in more intensive systems. These systems, or appropriate modifications of them, spread to other countries where there was a demand for poultry. In the oil-rich countries of the Middle East, the technology, the poultry stocks and the feed could be imported and used in intensive poultry production to add significantly to the diet, with minimal effort. With this explosion in intensification of poultry came an awareness that disease could destroy enterprises in which there were very large financial investments. There was a consequent development of interest by the veterinary profession. Today the success of the industry is dependent on disease control.

Poultry suffer from many diseases but there are a number of important areas in which disease control is particularly important for intensively managed poultry flocks.

These include:

the diseases of the alimentary tract;
respiratory diseases;
the Marek's disease – leucosis complex;
diseases affecting egg quality and number;
diseases resulting in immunosuppression.

Diseases of the alimentary tract

The diseases of the alimentary tract are extremely important. Coccidiosis of chickens is thought to be the most important disease of the poultry industry. Resulting from infection by protozoa of the genus *Eimeria*, economic loss is caused by the death or ill-thrift of young chicks. Birds that survive early infection become carriers and void the organism to provide fresh infection. Some *Eimeria* species may also cause overt disease in adult fowls. The infection is not transmitted to humans. Transmission to chicks is by way of cysts picked up from the litter in the brooding or rearing areas.

Control of the disease depends on knowledge derived from the work of parasitologists, immunologists, the drug industry and the practical poultry farmer. An initial factor affecting the course of the disease is the viability of the oocysts and their survival in the soil or litter. Very large numbers of propagules are excreted by infected birds in the oocyst stage. These oocysts are not infective but after a period of maturation in warm moist conditions (25–33°C) they produce sporulated oocysts in as short a time as two days. These spores are then relatively resistant and long-lived (up to two years in temperate outdoor conditions). In deep-litter conditions, however, the temperature generated, the ammonia released and the dryness in well-managed litter leads to the destruction of sporulated oocysts within about two weeks. The fact that immunity is developed by birds that are subjected to relatively low doses of the parasite (this immunity prevents later severe manifestations of disease) is also used in control. It is possible to seed the litter with low doses of the parasite but the preferred method is to give a coccidiostat at low dose in the food for a continuous period, during which time the bird is exposed to a relatively low level of infection and develops immunity, which it carries with it into adult life. The administration of the coccidiostat is usually withdrawn after about 12–16 weeks of age. At this point the birds should have developed immunity and should show no further signs of the disease. However, the skill of the poultryman is required to keep a close watch for any appearance of disease which might be associated with a local deterioration in the quality of the litter, due perhaps to a leak in a water point, allowing a buildup of an increased quantity of infective material.

Outbreaks of coccidiosis can be treated with sulphonamides, pyrimidines or amprolium. These drugs can also be used as, or incorporated in, coccidiostats but there are also a large number of disparate chemicals used, such as monensin, stenorol, clopidol, robenidine, and several others besides (Sainsbury, 1980).

Bacillary white diarrhoea caused by *Salmonella pullorum* was the most important cause of economic loss in poultry at the beginning of our period. It is not a direct threat to human health. In the late 1920s and early 1930s a critical situation arose, with many poultry keepers going out of business because of the death rate in their stock from this disease. *S. pullorum* (together with *S. gallinarum*) is different from most other *Salmonella* spp. in being restricted to birds, particularly poultry, and in being unable to colonise the alimentary tract directly. Their presence in the alimentary tract always follows previous tissue infection. They are thus able to cause the development of antibodies in the chicken; these antibodies can be identified by a simple blood test. Blood testing has been used to identify and eliminate the disease in breeding flocks. A private initiative by a group of growers in south-west England in 1932 became the voluntary Accredited Poultry Breeding Stations Scheme and in 1935 was transformed into the official Accredited Hatcheries Scheme, under the aegis of the English Ministry of Agriculture. In Scotland, the

Scottish Board of Agriculture had implemented a testing scheme with the help of the Royal (Dick) Veterinary College from 1912 to 1937, under which a bonus was paid to owners of stations approved for the distribution of eggs and day-old chicks; part of this scheme required health checks for *S. pullorum*. The result of the English scheme was a dramatic reduction in the amount of infection, with only 55 birds reacting positively out of a total of 4 283 146 birds tested in 1967. Claims have been made that Scotland appears to have been totally free of the disease since 1963 (Smith, 1971).

S. pullorum and *S. gallinarum* are not only antigenically related but also both show transmission through the infected oviduct to the egg and the next generation. In this they are different from other *Salmonella* spp. (about a dozen) which affect poultry. Of these the commonest is *Salmonella typhimurium*. None of these causes great losses in chicks, except in exceptional circumstances. They are imported into the flock from outside, from other chickens, faeces of carrier animals, eggs contaminated on the shells with faeces or infected litter, litter, etc., but apparently never from eggs with an internal infection, except in the case of ducks and to a lesser extent turkeys where there is some evidence of limited ovarian transmission. In general, infected chickens rid themselves of infection after a month or two, but a small percentage continue to excrete the organism for a longer period. *Salmonella enteridis* was becoming a problem in broiler flocks towards the end of our period; it could cause human food poisoning from infected broilers and could be transmitted through the eggs. Once identified it was possible to take appropriate steps to control it.

Respiratory diseases

The respiratory diseases are very important under intensive conditions. They include the virus diseases infectious bronchitis, Newcastle disease and infectious laryngotracheitis and the mycoplasma infection chronic respiratory disease. The virus diseases are controlled by the use of vaccines, which may be introduced in the drinking water, as a spray or by injection. The first two of these methods are particularly convenient for large scale use. Newcastle disease in its acute form was one component of the fowl pest disease syndrome (the other was fowl plague). Fowl plague was recognised in Britain in 1922, 1959 (in Scotland) and 1963 and each time has been eradicated. Newcastle disease was discovered by Doyle in 1927, in an outbreak in Newcastle. Other outbreaks occurred in Essex in 1970 and in pigeons in Britain in 1984. It has been the subject of a strenuous eradication campaign which, with the development of subacute strains and the appearance of effective vaccines, was discontinued in 1963 except for the peracute form and for all types in Scotland, which remains free. The disease is notifiable; the Ministry retained the right to institute slaughter in eradication campaigns up to 1986. The chronic respiratory disease caused by *Mycoplasma gallisepticum* cannot be controlled by vaccination but it can be eliminated by blood testing and careful selection. Most breeding flocks in the U.K., U.S.A. and Western Europe are now free of the disease.

Avian tumour viruses

The avian tumour viruses are represented in the U.K. principally by Marek's disease and leucosis. These two diseases were initially lumped together and confused under the name avian leucosis complex; measurements of economic loss mostly refer to this complex. Biggs (1971) considered that these viruses were responsible for about one third of the total annual loss from poultry disease in Great Britain and in money terms they represented a loss equivalent to about 5% of the total annual value of the industry. By 1971 he was able to say that the two diseases were readily distinguishable, and that Marek's disease had considerably more importance for the industry than leucosis and was the most important single cause of mortality in the poultry industry.

The control of these diseases has been achieved by a classic scientific study beginning with the induction of leucoses and of sarcomas by Ellerman and Bang in 1908 and by Rous in 1911 (Burmester, 1971). It was the first demonstration of the relationship between tumour production and virus infection. This was followed by the recognition of the disease complex in chickens with careful work finally distinguishing the thylaxovirus (with its major genetic component of ribonucleic acid) of the leucosis from the cell-associated herpes-type virus responsible for Marek's disease. Both viruses produce lymphoid neoplasia with a superficial similarity, which led to their initial confusion. The separation of the two diseases resulted from careful pathological histology, close study of the development of the diseases and eventual isolation of the viruses. In Marek's disease it was at first a mystery how a cell-associated virus could be so easily spread until the discovery of a concentration of the virus in the feather follicles and the demonstration of the release of the infective virus in dander and poultry dust.

This outline presents a rudimentary picture of the intensive research work that had brought about the control of Marek's disease, by the end of our period, using both vaccination of day-old chicks (to give life-time immunity) and careful isolation of the flocks under strict hygienic conditions. Vaccination for leucosis has not been successful but the disease is less easily spread than Marek's disease and it is possible to eliminate it from flocks. This is now undertaken by the breeding companies using ELISA (enzyme-linked immunosorbent assay) techniques for flock testing and eradication of the disease.

In both diseases there is evidence of genetic resistance in chickens, which responds to selection, but there is considerable difficulty in maintaining such a system of selection when all other measures are designed to keep the flocks disease-free.

Welfare aspects

The successful development of intensive poultry husbandry has been responsible for a great increase in the standard of living of the more developed parts of the world. Poultry meat, formerly a luxury, is now part of the general diet, and eggs are cheap and of high quality. On the other side of the coin have been the welfare problems developed by caging and crowding of egg-laying flocks and the unnatural and crowded conditions of deep-litter broiler production. It is also a matter of debate whether the methods used for transporting and finally slaughtering the broilers (and laying hens at the end of their useful life) are tolerable. Research work to analyse normal behaviour in domesticated animals and to seek a measurement of the stress developed in the agricultural environment has not yet provided answers to all of these problems but it has established parameters for unacceptable conditions, particularly for the most intensively confined livestock, namely poultry and pigs. In the case of the former much attention has been given to the design of battery cages and of improved systems such as the aviary system. In poultry the aim has been to provide secure accommodation which causes the least possible stress to the bird and retains the economic advantage of the system (Wood-Gush, 1983). In pigs, similar aims have been pursued with the additional aim of enriching the environment to reduce boredom and agonistic behaviour (Stolba and Wood-Gush, 1981). The work of the welfare societies (e.g. the Society for the Prevention of Cruelty to Animals and the Farm Animal Welfare Council) has placed the agricultural industry under pressure to change and has provided the stimulus for increased research work. Much requires to be done but at least the period of the second agricultural revolution did not end without the welfare consequence of the advances in applied science being identified and tackled in a scientific fashion.

Selected references and further reading

Aitken, I.D. (1983). Enzootic (chlamydial) abortion. In *Diseases of Sheep*, ed. W.B. Martin, pp. 119–23. Blackwell Scientific Publications, Oxford.

Armour, J. (1983). Control of gastrointestinal helminthiasis. In *Diseases of Sheep*, ed. W.B. Martin, pp. 250–4. Blackwell Scientific Publications, Oxford.

Barlow, R.M. (1991). Bovine spongiform encephalopathy. *Biologist* **38**, 60–2.

Barnes, E.M. and Mead, G.C. (1971). Clostridia and salmonellae in poultry processing. In *Poultry Disease and the World Economy*, ed. R.F. Gordon and B.M. Freeman, pp. 47–63. Published for British Poultry Science Ltd. by Longmans Group, Edinburgh.

Biggs, P.M. (1971). Marek's disease – recent advances. In *Poultry Disease and the World Economy*, ed. R.F. Gordon and B.M. Freeman, pp. 121–33. Published for British Poultry Science Ltd. by Longmans Group, Edinburgh.

Black, W. (1960). Control of *Nematodirus* disease by grassland management.

Proceedings of the 8th International Grassland Congress (Reading, Berks., 11–21 July, 1960), pp. 723–6. Published by the International Grassland Congress: Reading, UK.

Burmester, B.R. (1971). Viruses of the leucosis-sarcoma group. In *Poultry Disease and the World Economy*, ed. R.F. Gordon and B.M. Freeman, pp. 135–52. Published for British Poultry Science Ltd. by Longmans Group, Edinburgh.

Burns, K.N. (1981). *Animal Diseases*. In *Agricultural Research 1931-1981*, ed. G.W. Cooke, pp. 255–76. Agricultural Research Council, London.

Buxton, D. (1983). Clostridial diseases. In *Diseases of Sheep*, ed. W.B. Martin, pp. 35–43. Blackwell Scientific Publications, Oxford.

Buxton, D. (1983). Toxoplasmosis. In *Diseases of Sheep*, ed. W.B. Martin, pp. 124–9. Blackwell Scientific Publications, Oxford.

Cockrill, W.R. (1971). Economic loss from poultry disease. In *Poultry Disease and the World Economy*, ed. R.F. Gordon and B.M. Freeman, pp. 3–24. Published for British Poultry Science Ltd. by Longmans Group, Edinburgh.

Coop, R.L. and Christie, M.G. (1983). Parasitic gastroenteritis. In *Diseases of Sheep*, ed. W.B. Martin, pp. 56–61. Blackwell Scientific Publications, Oxford.

Dickinson, A.G., Stamp, J.T. and Renwick, C.C. (1974). Maternal and lateral transmission of scrapie in sheep. *Journal of Comparative Pathology and Therapeutics* **84**, 19–25.

Ely, F. and Petersen, W.E. (1941). Factors involved in the ejection of milk. *Journal of Dairy Science* **24**, 211–23.

Farm Animal Welfare Council (1984). *Report on the Welfare of Livestock (Red Meat Animals) at the Time of Slaughter*. Reference Book 248. Her Majesty's Stationery Office, London.

Frenkel, J.K. (1971). Toxoplasmosis: mechanisms of infection, laboratory diagnosis and management. *Current Topics in Pathology* **54**, 28–75.

Goodwin, R.F. and Whittlestone, P. (1967). The detection of enzootic pneumonia in pig herds. *Veterinary Record* **81**, 643–7.

Hobbs, B.C. (1971). Food poisoning from poultry. In *Poultry Disease and the World Economy*, ed. R.F. Gordon and B.M. Freeman, pp. 65–80. Published for British Poultry Science Ltd. by Longmans Group, Edinburgh.

Hope, H. (Ed.) (1994). Focus on Cattle supplement. *Farmers Weekly*, 18th March, 1994.

Knowles, N.R. (1971). Salmonellae in eggs and egg products. In *Poultry Disease and the World Economy*, ed. R.F. Gordon and B.M. Freeman, pp. 81–90. Published for British Poultry Science Ltd. by Longmans Group, Edinburgh.

Mathieson, A.O. (1983). Flock management for health. In *Diseases of Sheep*, ed. W.B. Martin, pp. 245–50. Blackwell Scientific Publications, Oxford.

Mitchell, B. and Stamp, J.T. (1983). Scrapie. In *Diseases of Sheep*, ed. W.B. Martin, pp. 71–5. Blackwell Scientific Publications, Oxford.

Ollerenshaw, C.B. (1959). The ecology of the liver fluke (*Fasciola hepatica*). *Veterinary Record* **71**, 957–65.

Pearce, J.W.R., Pugh, L.P. and Ritchie, J. (1965). (Ministry of Agriculture, Fisheries and Food.) *Animal health, a centenary 1865-1965*. Her Majesty's Stationery Office, London.

Robb, J. (1988). Mycotoxicosis: some facts and figures. *Pig Veterinary Journal* **22**, 75–82.

Rutter, W. (1975). Sheep from grass. *Bulletin No. 13*, East of Scotland College of Agriculture, Edinburgh.

Sainsbury, D. (1980). *Poultry Health and Management*. Granada Publications, London.

Sharp, J.M. and Martin, W.B. (1983). Chronic respiratory virus infections. In *Diseases of Sheep*, ed. W.B. Martin, pp. 12–17. Blackwell Scientific Publications, Oxford.

Smith, H.W. (1971). The epizootiology of salmonella infection in poultry. In *Poultry Disease and the World Economy*, ed. R.F. Gordon and B.M. Freeman, pp. 37–46. Published for British Poultry Science Ltd. by Longmans Group, Edinburgh.

Spedding, C.R.W. (1965). *Sheep Production and Grazing Management*. Baillière, Tindall and Cox, London.

Stevens, A.J. (1971). Economic loss from poultry disease in Great Britain. In *Poultry Disease and the World Economy*, ed. R.F. Gordon and B.M. Freeman, pp. 25–34. Published for British Poultry Science Ltd. by Longmans Group, Edinburgh.

Sterne, M. (1981). Clostridial infections. *British Veterinary Journal* 137, 443–54.

Stolba, A. and Wood-Gush, D.G.M. (1981). The assessment of behavioural needs of pigs under free-range and confined conditions. *Applied Animal Ethology* 8, 583–4.

Taylor, D.J. (1979). *Pig Diseases*. D.J. Taylor, 31 North Birbiston Road, Glasgow G65 7LX, UK.

Thiel, C.C. and Dodd, F.H. (1977, 1979). *Machine Milking*. National Institute for Research in Dairying, Shinfield, Reading.

Twiehaus, M.J. and Underdahl, N.R. (1970). Control and elimination of swine disease through repopulation with specific-pathogen-free (S.P.F.) stock. In *Diseases of Swine*, ed. H.W. Dunne, pp. 1096–110. Iowa State University Press, Ames, Iowa.

Whitelaw, A. (1983). Liverfluke. In *Diseases of Sheep*, ed. W.B. Martin, pp. 62–7. Blackwell Scientific Publications, Oxford.

Wood-Gush, D.G.M. (1983). *Elements of Ethology*. Chapman and Hall, London and New York.

12 Integrations in animal husbandry

As the pace of agricultural change gathered momentum, so new ways of farming the land associated with animal enterprises came into being. Almost all involved the intensification of production methods and many, indeed with pigs and poultry virtually all, resulted in the importation of feed for stock. Large lorries transporting processed feed from the mills and extrusion plants of the feeding-stuffs firms became commonplace on country roads. Both large and small farmers also purchased feeds which they then mixed themselves. All farmers had ready advice at hand when they changed to new methods. Not only did the advisory services of government provide this free of charge but the representatives of the feed firms also provided guidance, albeit not quite so dispassionate. The farming press helped with informative articles and with stories of the successes of individual farmers. Many local farmers clubbed together to form local discussion groups (often with the local adviser as secretary and unobtrusive guide) to which experts were invited to extol the virtue of some innovation and where individual experiences and indeed prejudices were well ventilated.

The period from the 1950s to the 1970s was the main one in which this massive transfer of technology from research, through development and into practice took place. Adoption of new methods often led to the emergence of new problems and certainly to a demand for the rationalisation of the accrued knowledge that had resulted in their genesis. Nowhere was this more true than in the provision of advice about the adequacy of the diets which were being given to stock; the feeding-stuff firms, the advisory chemists of government, and increasingly the farmers themselves, needed far more information on which to plan the feeding of farm animals and resolve any immediate difficulties that ensued. In 1959 the Agricultural Research Council established a Technical Committee of scientists under the chairmanship of Sir David Cuthbertson to undertake the task of rationalising the existing information and estimating the nutrient needs of all classes of farm livestock. This work was based on the many thousands of research papers that had been published throughout the world and was somewhat analogous to that undertaken by Crowther and Yates in the 1940s when the results of thousands of fertiliser trials were summarised to provide rational recommendations for the manuring of crops (Chapters 3 and 5). The animal studies were of considerable importance in other ways. The results provided entirely new concepts about approaches to nutrient needs of livestock and these, first formalised in Britain, besides aiding British livestock farmers, have become the basis for livestock feeding in most other countries in Europe. Two examples of these new concepts relating to ruminant animals and one relating to pigs are adumbrated below. In addition some examples are presented of the new technologies of livestock husbandry which emerged and the new and often unexpected problems (demanding research) that they uncovered.

Nutritional requirements

Ruminant needs for energy from feed

Ruminants can do what species with simpler digestive tracts cannot: they can digest and use the energy from the structural carbohydrates of plants, which include hemicellulose and cellulose (the most plentiful organic molecules in the world). This ability is conferred by the microorganisms in the capacious rumens of these animals. A century ago animal physiology was dominated by a concern with human physiology and it was thought that ruminants obtained their energy in much the same way as humans, namely as simple sugars after digestion of starch. This view was reinforced by the experiments made by Gustav Kuhn and Oskar Kellner, the foremost agricultural chemists in Germany at that time. They compared starch with purified cellulose in experiments in calorimeters to find that when equal amounts were fed the cattle stored much the same amount of energy in their bodies. Starch and purified cellulose were obviously equal sources of energy, and following Kellner, for many years rations were calculated in terms of starch equivalents. It was of course realised that the microbes fermented the fibrous material the animals consumed but it was equally thought that the end product of the fermentation must be a simple sugar. This view was overthrown in 1944 in the laboratories of the Physiology Department at Cambridge University by Sir Joseph Barcroft, A.T. Phillipson and Rachel McAnally. The end products of the fermentation were found to be the lower volatile fatty acids (VFAs) acetic, propionic and butyric acids (which were separated by steam distillation). The fact that the nutritive value of starch was close to that of the purified fibre in Kellner's experiments was because both were fermented in the rumen to give these volatile fatty acids.

These findings were the most seminal ever to be made in the study of ruminant metabolism. Sir Joseph Barcroft was a distinguished physiologist who had illuminated many areas including respiration, the role of haemoglobin, and latterly the metabolism of foetal life. A.T. Phillipson, who worked with him, was a veterinarian who moved to the Rowett Research Institute and finally became the first head of the newly established Veterinary School at

Cambridge. It was a most fruitful collaboration, which engendered a great amount of further work throughout the world.

Methods for the separation, analysis and identification of the VFAs were devised by S.R. Elsden. He later became the Director of the Food Research Institute in Norwich. The isolation and characterisation of the hundreds of species of bacteria involved in ruminal fermentation was undertaken by P.N. Hobson of Aberdeen and R.E. Hungate of the University of California, at Davis, while among the other components of the system, the protozoa were investigated by J.A.B. Smith (who later directed the Hannah Research Institute) and G.S. Coleman of the Physiology Department of Cambridge University. The fungi were also important constituents of the rumen flora. What emerged was the incredible complexity of the ecosystem of the rumen. Earlier it had been hoped that it would be possible to manipulate the flora and fauna to improve its ability to digest cellulosic material, but this was not achieved. The flora was an open one, modified by diet it is true, but resistant to perturbation. One exception has been found to this. The leguminous shrub genus *Leucaena* contains a highly toxic amino acid, mimosine, a hydrolysis product of which causes goitre in stock that graze it. It was found that goats in Hawaii were able to exist on this toxic material because their rumens harboured an organism that was capable of degrading the toxic material further. R.J. Jones and R.G. Megarrity, working in Queensland, Australia, showed in 1986 that the Hawaiian organism could be transferred to and established in the rumens of cattle in Australia, enabling them to thrive on this otherwise toxic feed; this finding has opened up vast areas of subtropical lands to this useful legume.

Metabolic studies on the fate of the volatile fatty acids followed their discovery to widen still further knowledge about the role of the ruminal flora. G.A. Garton discovered that fats (triacylglycerols) – to which ruminants have low tolerance – are completely hydrolysed in the rumen and their constituent unsaturated fatty acids hydrogenated. This implied that the absorption of fat by the ruminant must follow a different course from that in animals with simple stomachs, and so it proved. Surgical methods were developed by J.L. Linzell at the Institute of Animal Physiology at Cambridge for the measurement of the export of the products (VFAs) of the rumen fermentation from the gut and of their import by the mammary gland. These techniques, coupled with the use of radioisotopes of the volatile fatty acids, first by E.F. Annison and D.B. Lindsay and subsequently by R.A. Leng in Australia, provided quantitative measurements of the rates at which the volatile fatty acids were produced and, in confirmation of the work of S.J. Folley and G. Popjak at the National Institute for Research in Dairying at Reading, how they contributed to the synthesis of the constituents of milk. The effect of the nature of the diet on the proportional makeup of the mixture of volatile fatty acids in the fluid of the rumen was examined in many laboratories to show that, although acetic acid was invariably present in highest concentration, starchy and sugary foods favoured the production of propionic acid. Emulating the experiments of Kellner 60 years before, D.G. Armstrong, K.L. Blaxter and N.McC. Graham in Scotland used calorimetric methods to measure the nutritive value of the volatile fatty acids to reveal that as a source of energy they were inferior to glucose (which was given intravenously to avoid ruminal fermentation) and that propionic acid was superior to acetic and *n*-butyric acid in terms of promoting the deposition of fat and protein in the body. Other work had shown that diets promoting propionic acid production are more efficient as there are reduced losses of methane (a point exploited by certain growth-promoting substances such as monensin). This work linked the biochemical findings to the metabolism and energetics of the whole animal.

The growth of knowledge about the digestive physiology of ruminants was immense during the 1950s and 1960s; much of the new knowledge had application to specific problems. The low concentrations of fat in milk produced by cows consuming diets containing large proportions of grain was shown by C.L. McClymont in Australia to be, in effect, a deficit of acetic acid; indigestion was shown to derive from the accumulation of lactic acid; and bloat – a condition in which the rumen becomes distended by gas – was related to a series of microbial and feed effects combined. Undoubtedly the most important implication of the new appreciation of ruminant metabolism, however, was to sow discontent about the existing ways in which the energy requirements of farm livestock were assessed. These were based on the starch equivalent system, which had been devised by Kellner in Germany at the turn of the century and were used throughout Europe. They were applied to all classes of stock – ruminants, pigs and poultry alike – and were highly unsatisfactory, for they did not explain why some feeds were superior to others or why some classes of stock did poorly when rationed according to the system.

In 1962, and following about ten years of intensive calorimetric work at the Hannah Research Institute at Ayr, Scotland, K.L. Blaxter proposed a new system for meeting the energy needs of ruminants in which cognisance was taken of the fact that the energy provided by a feed was not constant but varied according to the type of diet in which it was incorporated, the total amount of it that was given and the function – such as maintenance of weight, growth, fattening, or milk secretion – that it supported. This new system, called the metabolisable energy system or the A.R.C. system, was incorporated in the report on ruminants by Sir David Cuthbertson's committee and a little later was adopted by the government departments of agriculture to aid its rapid dissemination. Geoffrey Alderman, the senior agricultural chemist of the Ministry of Agriculture, ran courses for advisors to familiarise them with what was a more complicated scheme than that with which they had struggled for so long. In addition, feed evaluation units were set up in England and Scotland to provide the basic data on the nutritive value of feeds necessary to implement the new system. Critical here was the need for good analytical methods to describe feeds, and particularly the more fibrous ones on which ruminants largely depend. The analytical scheme then in use throughout the world was one devised in Germany in the middle of the nineteenth century by W. Henneberg and F. Stohmann, called the 'Weende system'. Everyone knew it was unsatisfactory but there was reluctance to change, largely because there had been so much information accumulated using it. Methods were devised in the 1930s and 1940s to estimate particular components of plant cell walls, but the major change took place when P. van Soest of Cornell University in the 1960s devised an analytical scheme based on the use of detergents to separate components in a more meaningful way. At the Grassland Research Institute, Hurley, England, in 1963, J.M.A. Terry and R.A. Terry derived an analytical procedure to estimate the digestibility of feeds by ruminants. This entailed incubating the feed sample with ruminal fluid containing the anaerobic organisms followed by treatment with enzymes from the gut. The correlation between what ruminants actually digested and this *in vitro* estimate was very good.

The metabolisable energy system was adopted in most countries in Europe and in the (then) Soviet Union, modified in some instances to meet local conditions, and recently has been adopted with further modification in Australia. More recently proposals have been made for a parallel metabolisable protein system; and the basic principles on which this could be based are discussed in the following section.

Ruminant needs for protein

It had been known since the latter half of the nineteenth century that ruminants could make use of non-protein nitrogen-containing materials, such as urea or the many simple nitrogenous compounds present in herbage or other feeds, as precursors of protein for growth. No-one knew how much they could use, and so the unit in which the protein requirements of animals was assessed was in terms of the 'protein equivalent'. In this unit, and quite arbitrarily, non-protein nitrogen-containing materials were given half the value of the protein. Such a unit of nutritive worth was hardly suited to the demands of a sophisticated animal industry.

As the new knowledge of the ruminal fermentation processes spread it was realised that much, but not necessarily all dietary protein was hydrolysed in the rumen to give rise to amino acids and that these were then deaminated to give rise to ammonia. As was demonstrated by J.A.B. Smith, following careful experimentation, urea was also subject to degradation by potent microbial ureases to form ammonia. I.W. McDonald provided an integration of the information that was accruing. He showed that although ammonia could be absorbed directly from the rumen, for the most part it was used to support microbial protein synthesis. The microbes, together with any dietary protein that had escaped fermentation, passed on through the gut, where they were digested by alimentary tract enzymes to provide a large proportion of the amino acids needed to support the growth of the body. Moreover, the microbial protein was of high biological quality. To meet the protein needs of ruminants, therefore, the microflora and microfauna had first to be fed before considering the animal proper; dietary systems had also to take account of ruminant energy–protein interactions.

It was realised that these concepts could form the basis for a system of practical feeding. To do so, however, entailed providing quantitative information about every aspect of these complex processes. An immense amount of work followed. This involved estimating the quantitative flow of microbial protein from the rumen. Surgical techniques that had first been developed by Russian physiologists were refined by Phillipson and J. Hogan to create re-entrant cannulas so as to measure the rate at which digesta moved. The re-entrant cannula provided an artificial bypass of particular portions of the gut on the external surface of the animal, thus allowing the digesting food to be sampled as it moved through the body. The animals did not appear to be upset by this interference and generally lived for a considerable time without complications. These surgical techniques were successfully applied by Hogan on his return to Australia, by D.G. Armstrong at the University of Newcastle, by R.H. Smith at the National Institute for Research in Dairying at Reading, and by others. Methods for distinguishing microbial from non-microbial protein were developed, based on the presence of unique amino acids in the cell walls of the microorganisms or, alternatively, by labelling the microorganisms with radioactive sulphur. The effort was international. In the early 1970s schemes for meeting protein needs were developed in the United States of America by W.S. Burroughs and his collaborators; a similar scheme was developed in Germany by W. Kaufmann and H. Hagemeister. The group of British workers announced their scheme in 1972; it was a more complicated approach than that of the Americans or the Germans and was adopted. It solved many problems related to how best to make use of feeds as protein sources. The American system was called the 'Metabolisable Protein System' (to match the Metabolisable Energy System); and the British system was based on 'rumen degradable and undegradable protein' to reflect the fact that only a proportion of the protein escapes ruminal digestion and is digested and finally absorbed in the small intestine (McDonald, Edwards and Greenhalgh, 1981).

The nutrient needs of pigs

The major problem encountered when attempting to rationalise the accumulated information on the nutrient needs of pigs was that yesterday's pig was not the same as today's pig and tomorrow's pig was likely to be different again. Genetic improvement and breed changes had created a different pig characterised by a very high rate of growth of the lean tissues of the body. Much of the older information on fatter and slower-growing pigs was thus irrelevant in a quantitative sense. In addition there was an emphasis on the use of plant sources of protein, which were replacing the almost obligatory 5–7% of fish meal included in the diets of pigs before the Second World War. The carbohydrate source, too, was changing, with an increased emphasis on barley and a diminution of the amounts of imported maize incorporated in the pig's rations.

Initial attempts to arrive at precise estimates of the protein requirements of pigs were complicated by the fact that not all proteins were equal: they contained different proportions of different amino acids. With barley-based diets the level of the amino acid lysine was the first to limit growth (Chapter 7) and some resolution was obtained by taking it into account. However, not all the lysine measured chemically in a feed appeared to be useful to the pig; pigs did not always grow well when given diets apparently amply provided with lysine. Studies of the deterioration of milk powders during storage, by Kathleen Henry and S.K. Kon at the National Institute for Research in Dairying in 1950, had shown that non-enzymic browning of the powders reduced the biological value of the milk protein (a measure of the overall amino acid adequacy) by 20%. This browning was a chemical reaction between the epsilon amino group of the lysine and the lactose in the powder. Equivalent changes took place with other carbohydrates in feeds, similarly rendering their lysine unavailable. Gabrielle Ellinger and K.J. Carpenter in 1955 devised an analytical method to distinguish the unbound or 'available' lysine and this method resolved many of the anomalies found in feeding tests with both pigs and chickens. The method was adopted world-wide. Japanese companies had produced synthetic lysine by a fermentation process and the amino acid was, and is, extensively added to the diets of pigs.

The problem remained: how to define the protein requirements of pigs more precisely. The ideal solution would be to define, not in terms of protein at all, but rather in terms of essential and non-essential amino acids. Several groups of workers throughout the world attempted this. It was a difficult task because the dose–response relationships were very flat (i.e. little change in response to different doses) and experimental methods to determine optima called for great care. Many of the technical constraints were removed by M.F. Fuller of the Rowett Research Institute at Aberdeen, who was able to obtain good estimates of the essential amino acid needs of pigs, as was D.J.A. Cole of Nottingham University (1978). These findings were incorporated in 1980 into a definition of an ideal protein with the amino acid composition that entailed no waste by over-supply and no impairment of growth by under-supply. Further work by Fuller refined these estimates. Definition in this way means that the protein requirements of pigs growing at different rates – and hence retaining in their bodies different amounts of protein – can be expressed in absolute terms of an ideal protein, that is, in terms of amino acids. The ability of a particular feed protein to meet these requirements can then be assessed explicitly in terms of whether it conforms to the composition of the ideal protein.

Science into practice

Many of those who are concerned with agricultural science have a considerable agricultural awareness and can see ways in which practical application of their own work or that of their fellows can be made in complete farm systems of husbandry and management. Some of these applications involve considerable investigational work, the testing of alternatives and the exploration of consequences; the work is not confined to farm tests but may also require specific pieces of pure research to help the overall resolution of the problem. Some examples of these attempts to change animal production systems and the factors that led to their success – and in some instances failure – are given below.

The partition problem

In the case of the dairy cow, feeding standards were somewhat static whether based on the old starch equivalent system which held sway between the two world wars or the new metabolisable energy system, put in place in the 1960s. The amount of feed to be supplied was simply dictated by (for example) the amount of milk the cow produced. The standard took no account of the fact that milk production responds to additional feed and is depressed by insufficiency. Not only does milk yield increase when more feed is given but so too does the body weight of the cow. The cow partitions increases in food which she is given by allocating some to increasing her milk production and some to increasing her body weight; this partition varies over the lactation. Early in her lactation she loses body weight to support milk production; later, body weight gains tend to become more important. These latter changes were vividly illustrated by W. Flatt and P. Moe in their calorimetric experiments at the U.S.D.A. centre at Beltsville, Maryland, using a very high-yielding cow, 'Lorna'.

It was largely as the result of an interest by R.A. Fisher that a programme of investigation into the efficiency of utilisation of feed by dairy cows was commenced at the Hannah Research Institute by K.L. Blaxter and H. Rubin. The results showed that there was a strong historical component to the course of lactation and that responses to increases or decreases in the feed supply were not only delayed but could be predicted so as to arrive at economically optimal inputs for different cost and price structures. This work was taken much further by W. Broster at the National Institute for Research in Dairying in an important series of investigations to examine how best to allocate food over the whole lactation and even between lactations. The practical outcomes of this work were simplified systems of dairy cow feeding, which were well adapted to the management of the very large herds – over 300 cows in some instances – that developed within the latter part of the 50-year period. Within these systems the nutrient requirements were, of course, still met through appeal to the feeding standards: diets were still designed to be nutritively adequate but recognition was given to the fact that milk output is not only an intrinsic attribute of the cow but is also sensitive to long-term nutrition.

Intensive beef production

At the beginning of our revolutionary period, the conventional methods of beef production entailed the use of both summer grazing and winter fattening in yards of beef-type animals to produce a finished beast at the end of about two years. In the late 1950s R. Preston of the Rowett Research Institute realised that, with the expansion of the national dairy herd and the change of its breed structure towards Friesian cattle, there were calves, some sired by beef breeds but many of dairy conformation, that would undoubtedly provide much of the beef in the future. In addition these calves were then relatively cheap. So, too, was barley, as the new varieties and technologies of cereal cultivation were adopted. Preston also realised that calorimetric work had shown the high value of barley when given as the main source of energy for the ruminant. After many attempts Preston devised a method of processing barley by lightly rolling it and mixing it with a concentrated preparation of protein, minerals and vitamins to make a complete diet. Cattle consumed this avidly and on it they gained weight at a rate well in excess of 1 kg per day. A finished animal could thus be produced in ten months. The 'barley beef' system, as it was called, was taken up by many farmers, as it enabled continuous and intensive production to take place within one set of premises and was profitable. The proportion of United Kingdom beef produced in this way rose to over 10% of the total within a few years.

Preston eventually, at the invitation of Fidel Castro, moved to Cuba to establish a research facility there for the Cuban government. He has made the rest of his career in the Caribbean where, with the assistance of M. Willis, he has devised effective methods of beef production based on the use of waste products from the sugar industry.

There were, however, problems associated with barley beef. It was found that some cattle so fed developed abscesses in the liver. These were initially controlled by mixing antibiotics with the diet, but the primary cause resided in the effects of the diet on the ruminal wall. B.F. Fell showed that the hair that the animals consumed when they groomed themselves, together with the minute siliceous spines of the barley, became aligned with the papillae of the rumen wall, which were grossly distorted by the starchy diet. They caused penetrating wounds of the mucosa, which allowed pathogenic organisms to enter the portal blood stream and hence the liver. Alterations to the diet by the inclusion of bicarbonate and very small quantities of fibrous feed proved to prevent the condition. Bloat (see above) was a recurring problem and required careful management of the feeding regime.

The idea that cattle could be fed a complete diet that could be handled mechanically with consequent saving of labour was one which was taken up widely even if the extreme of all-grain feeding was for the specialist. The farm machinery industry responded by making special mixing and delivery trailers. Animal houses were redesigned to take these machines and to provide increased ventilation. Easier methods of slurry disposal through slatted floors were developed. Designs were also developed to ensure the safety of the operatives especially where the enhanced growth rate of bull beef was exploited, loose-housed young bulls being notoriously unreliable. The development of the systems using dairy (male) calves was of benefit to the dairy industry, which could now be assured of a ready market for male calves and, with a tight calving pattern, could contract with other farmers to deliver agreed numbers of calves at particular times of year.

Intensive feeding of sheep

It was inevitable that the work Preston had done with cattle would be extended to sheep with a view to the intensification of the production of lamb. The cattle studies were transferred to sheep very successfully by E.R. Orskov; excellent results in terms of growth rate were obtained, such that finished lambs could be sent to market about 80 days from birth. There were no problems due to damage to the ruminal wall since lambs are woolly; unlike cattle, they have no hairs to make wounds. However, an entirely new problem arose. When the carcases of these lambs cooled, their superficial fat did not solidify but remained soft. This led to the discovery by W. Duncan and G.A. Garton in 1972 of a hitherto unsuspected series of curious fatty acids in the lamb's body lipids. These were fatty acids which had single methyl branches, and some of them had uneven numbers of carbon atoms in the aliphatic chain. A series of experiments showed that what was happening was that the large amounts of propionic acid produced during the fermentation of the barley was replacing acetic acid in the subsequent biosynthesis of the fatty acids. More specifically, propionyl coenzyme A and methylmalonyl coenzyme A were replacing acetyl coenzyme A and malonyl coenzyme A in the fatty acid synthetase complex. A Vitamin B12 deficiency was suspected because Vitamin B12 is a component of the enzyme system that normally converts methyl-malonate to succinate and prevents its accumulating; however, supplements of the vitamin were without effect. It seemed that sheep, unlike cattle, could not cope with excess propionic acid. The biochemistry was fascinating; the practical problem was real. It was solved quite simply by substituting untreated barley – the whole grain – for the lightly rolled barley. The result was a sufficient fall in the production of propionic acid and an increase in acetic acid production by the ruminal flora to prevent the relative surplus of propionic acid.

With increased knowledge of the nutrition of sheep and methods available for increasing fecundity (by flushing (see below) and by genetic means (Chapter 10)) it became practicable to plan for a higher lambing rate by the production of twins by most of the ewes. Precision was given to this by the introduction of ultrasound scanning, which allowed twin-bearing ewes to be identified early in the pregnancy and their nutrition adjusted accordingly.

Methods of grazing control

Hundreds of replicated plot experiments to determine the ways in which to grow large yields of grass of high nutritive value had been conducted by the early 1950s. In Britain W. Holmes, then at the Hannah Research Institute, Ayr, and T.T. Treacher at the Grassland Research Institute, Hurley, led the way. These experiments were supplemented by determinations of the digestibility of the cut herbage by F. Raymond at Hurley and of its energy value by D.G. Armstrong at Ayr. The attributes of different species and cultivars, the role of different cutting frequencies in determining total yields in a season and the best sequences in which to apply mineral and nitrogenous fertiliser had been worked out for different climatic and weather conditions. Questions arose about how, on the basis of these results, the grass crop could best be used to support milk production, not in simple tests with single cows, but with whole herds so as to emulate real farm conditions. The crop appeared so valuable that one answer was to ration it by what was called 'strip grazing'. At each grazing, morning and evening, the cows were allowed a fixed amount of grass by placing an electric fence to restrict the area grazed to a strip, which they then consumed. Experiments in the United Kingdom usually showed that there was an advantage to this rationing system over what was called 'set stocking', that is allowing the cows access to the whole area to be grazed. Quite contrary results were obtained in New Zealand, where C.P. McMeekan, the Director of the Ruakura Research Institute, found no difference between the two systems of grazing management. McMeekan showed that the determinant of output of milk was simply the stocking rate: the number of animals that were allowed to graze. This is understandable, as pointed out by G.O. Mott, the American agronomist from Indiana. If there is an excess of herbage then any increase in the number of animals grazing an area will increase milk output. Conversely, if milk production is limited by shortage of herbage, then more stock grazed on it will result in a reduction in intake by each individual, with a consequent diminution in yield. It appears in retrospect that a subjective bias may well have been introduced into many of the experiments in the United Kingdom by unwittingly stocking the strips more heavily than the field. The results of these and many other experiments led to the adoption of paddock grazing in which cows were pastured on a series of larger areas which were then grazed, or given fertiliser, or conserved for silage in rotation. The electric fence still proved very useful in that it gave an additional means of control of the all-important stocking rate.

In the late 1970s and early 1980s a new approach was made to the technology of grazing. Instead of growing a prodigious amount of grass and then devising ways in which it could best be grazed to sustain an acceptable level of production, the view emerged that the production of grass that the cow was avid to consume would support a high level of production and that the objective should be to obtain the highest possible yield of such material; in other words, a qualitative measure was introduced and the cow was asked her opinion of what she liked! Much investigational work was done to achieve the integration of these different methods. W. Holmes at Wye College described the voluntary consumption patterns of dairy cows in differing circumstances, while R.J. Wilkins and grassland agronomists at Hurley, and J. Hodgson of the Hill Farming Research Institute at Edinburgh, were concerned with the dynamics of pasture growth. More practical studies were made by J. Leaver of the West of Scotland College of Agriculture at Ayr. What emerged was a concept of balance between the various processes going on during the growth and utilisation of the crop. These include not only the biosynthesis and growth of new material but its removal by the animal and by natural decay. Using optimising techniques it was found that to assure an ideal intake of herbage by the individual cow, and consequent ideal utilisation of the herbage grown, the grass must be so managed that it is between 7 and 10 cm high at the time of grazing. To maintain these ideal conditions for the cow, stocking rate is even more important, calling for the use of the electric fence to provide what is called 'buffer grazing'. This is used when the weather restricts growth or to reserve surplus grass for conservation when growth of grass exceeds the animals' needs.

The problem of hill grazing

The hill lands of Britain represent a large resource of poor intrinsic worth. About 30% are heather hills where grouse compete with sheep; the remainder consists of grassy hills where the herbage is derived from species of low nutritive value across the year. This is in part because, from their altitudinal situation, they yield a considerable weight of nutritional material over a very short period at midsummer and it is impossible to match this growth with an adequate stocking rate, leaving much senescent and indigestible material to accumulate from year to year. The harvesting of these large areas is by continuous grazing by hill breeds of sheep; the number grazed on any holding is determined by its ability to provide sufficient palatable feed during the winter. The under-utilisation of the summer herbage leads to ecological change and cycles of sward deterioration. A further difficulty is that hill farmers have few financial resources: obviously winter stocking could be increased by provision of supplementary feed bought in for the purpose, but this, in general, is not possible for the cost in relation to the returns is high.

In a remarkable programme of applied research, workers at the Hill Farming Research Organisation, led by J.M.M. Cunningham, J. Eadie, T.J. Maxwell and others, devised ways in which capital investment in hill land improvement could be paid for by increasing returns on that investment (Chapter 13). Sound biological principles were adopted in assessing the weak points in the existing system – and indeed those uncovered during the development of new ones – to arrive at what has been called 'the two pasture system'. In this system, relatively small areas of the whole of a hill farm are improved by reseeding them; some are fenced; some are seeded as a mosaic on the hill. These provide good-quality feed when the ewe needs it most, namely during late pregnancy and lactation. The fenced land can be efficiently managed by herding stock on and off at appropriate times. Surplus vegetation of digestible species which arises in midsummer can be utilised in the early part of winter. The system allows a greater production of better-quality lamb and sufficient income to support the capital expenditure. The critical feature, somewhat different from gross margin analysis described in Chapter 3, was to place emphasis on cash flow. Large-scale demonstrations on grassy hills and heather hills indicated that, after several years of operation of the system, output of lamb could be almost double that of the traditional systems of flock management and so too could income. The work was taken up by the advisory services in both England and Scotland and widely adopted by pioneer flockmasters.

Analogous methods were then applied to upland grazings. These are permanent pastures at lower elevation than true hill farms and called for different approaches in terms of land improvement, including the possible housing of sheep to spare pastures over winter. Here the principles were closer to those adopted in lowland grassland husbandry. There were, in addition, a number of new findings that also had application elsewhere. Notable were those by J. Gunn, who showed that a higher feed intake before mating ('flushing') increased fecundity, and by J. Hodgson who showed that, just as milk production in cows was best maintained if sward height was kept at 7–10 centimetres, so with sheep optimal output was obtained if herbage was kept at 4–6 centimetres and in certain circumstances less. To the eye of the uninitiated this sward height may give the impression of being a bare pasture in need of rest!

The conservation of fodder

The conservation of grass for winter feed had since time immemorial been as hay, and so it continued to be in the 1930s. Much of the hay crop was cut when the grasses and clovers were mature and highly fibrous, and of low digestibility. Early work had shown that the nutritive value of the nation's hay could be improved if only the farmers would cut the crop when it was green and leafy. Making hay from young grass is, however, difficult, particularly in wet seasons. In these circumstances the crop has to be turned frequently to allow access of air and sunshine to dry it and, as the crop dries, the valuable leaf tends to shatter and is lost while the longer drying process necessary for young grass results in a greater loss of dry matter through the continuing respiration of the still-living cells. Studies of the overall losses that occurred in making hay were made by numerous investigators; these were collated in 1939 by S.J. Watson to show that losses of 30% of the dry weight of the original crop were not uncommon. In the late 1930s workers at the Jealott's Hill research centre of Imperial Chemical Industries (I.C.I.) embarked on investigations to find whether it was possible to conserve young grass as silage so as to reduce losses, to have a conservation process more independent of the United Kingdom's fickle weather and to produce a winter feed of higher nutritive value than the poor hay that was so commonplace. Various aspects of the technology of silage making and of the problem of its establishment in the British agricultural industry are described in Chapters 8 and 13.

In the 1920s H.F. Woodman of the School of Agriculture at Cambridge had been undertaking experiments related to the nutritive value of grasses and clovers; he drew attention to the fact that the composition of the young leafy grass was close to that of imported protein-rich cakes and meals. Could drying of the crop on a farm scale be achieved, there was the possibility of a home-grown source of protein for all classes of stock. When S.J. Watson at Jealott's Hill was looking at alternatives to the hay crop he considered this possibility. I.C.I. employed engineers highly knowledgeable about the thermodynamics of drying at Billingham in Northumberland, and they built a succession of what were called the Billingham grass-driers, commencing in 1936. The machines were expensive in terms of capital cost and expensive to run in terms of fuel consumption. They were taken up by a very small number of farmers. Even these, with one notable exception, fell into disuse during the Second World War. The exception was at the Hannah Research Institute where N.C. Wright (the Director) and his staff used intensive methods of grassland husbandry, some artificial dehydration and judicious cropping to make the farm completely self-sufficient and independent of the amounts of feeding-stuffs that could be purchased at that time.

However, resurgence of interest in artificial drying emerged in the 1960s. Much work showed that land cropped with grass for artificial drying resulted in greater animal production per hectare than the alternatives. In alternative systems such as direct grazing there was the wastage of the herbage from treading and defecation; in silage making there was also wastage (losses were about 20%); and comparisons of dried grass with fodder root crops and barley showed that about double the meat output per hectare could be obtained. The problem was that the rate of gain per unit of time was lower, and time is money. The reason for the low rate of gain was that the high-quality material that Woodman had produced from spring grass could not be maintained over the season; later harvests were of poorer quality. Some farmers took up drying of grass, lucerne and other crops on a large scale but increasingly they catered for specialised markets where the carotenoids of the green material were highly valued. The oil crisis of the early 1970s effectively stopped further development.

The development of the poultry industry

The development of the egg-laying and broiler industries was a continuous process, and although there were phases, it can conveniently be dealt with as a whole. In the 1930s 98% of eggs were produced from birds kept on free-range or in semi-intensive systems. Forty years later, after a mild flirtation with deep litter houses in the 1950s, in 1977 93% were housed intensively in battery cages. The by-product of the laying flocks of the 1930s were the cockerels destined for table poultry; these were a luxury product. The pullets were separated from the cockerels by using the sex-linkage of colour, feather pattern or leg colour. This linkage had resulted from some very elegant genetical work by R.C. Punnett and M. Pease; at first they used established breeds (e.g. White Wyandotte and Brown Leghorn). Their work then resulted in the development of a new breed, the Cambar, and then in the transfer of appropriate genes into established breeds or hybrids. Alternatively the cockerels were distinguished by an examination of the cloaca of the young chick, at first by Japanese sexers and later by skilled operatives from the U.K. industry. By 1986 an entirely new industry had grown up, the broiler industry, where specially bred meat-producing birds (of both sexes) were produced in a vertically integrated system from hatcheries, through production plants, to processing and freezing plants and retail outlets. The time taken to produce a standard four-pound liveweight bird in the 1930s was about 125 days; by 1986 it was only 42. Poultry meat, far from being an occasional item of diet for special occasions, became a commonplace. As was shown in Table 7, egg production per bird in laying flocks increased by 80% in the same period.

These massive changes were dependent to a considerable extent on technologies developed in the United States and adapted by pioneers in the United Kingdom to suit local conditions. Nevertheless, real scientific and technical advances were made in the United Kingdom in nutrition and in disease control. The latter is dealt with in detail in Chapter 11.

The broiler industry in the United States is thought to have commenced on the Delaware peninsula in the 1920s. At that time conventional breeds were used, but by 1942 the first hybrid strains were introduced by H. Wallace of the

Hiline Poultry Breeding Company in the United States to meet the needs of the industry for a fast-growing bird. British interest was considerable, augmented by an official mission to the U.S.A. in 1946. The mission brought back many new ideas, including those related to broiler production. Nothing much could be done at that time to adapt the American system to British conditions because of the post-war shortage and poor quality of available feed-stuffs. In 1948, however, the first *legal* importation of American meat-line broilers was made by R. Chalmers-Watson of East Lothian, Scotland, to commence the industry. Many others followed, notably G. Sykes in 1953 and J.B. Eastwood in 1959. New techniques to aid the processing of birds were also invented and adopted. These included machines, mostly of American origin initially, for plucking the birds, and the blast freezer, which was devised by J. and E. Hall of Dartford. Links, too, were made with retail outlets; Messrs Marks and Spencer took the lead in 1959 and the scale of operations became immense. In 1960 one firm, Messrs Thornbers, sold 18 million day old chicks, and a few years later four major companies dominated the processing side of the industry. By 1974 the market was dominated by two companies, whose products stretched from the egg to the finished broiler carcase: J.B. Eastwood Ltd. and Ross Poultry Ltd. At that time J.B. Eastwood owned about 5 000 hectares and processed over a million broilers a week, simultaneously producing in their egg-laying plants about 18 million eggs. The egg-producing venture was the largest in the world.

These large enterprises drew on research from across the world, testing alternatives themselves and also undertaking their own investigations. They were also helped by the Agricultural Research Council's Poultry Research Centre at Edinburgh and by the government advisory service. New hybrid strains from the United States, Holland, Denmark and South Africa were introduced and the latest technologies of environmental control and nutrition adopted with alacrity. Similar techniques were used with ducks and turkeys to produce parallel industries. All these industries were far removed from what had been regarded as farm operations: they represented the industrialisation of animal production or, as it came to be known, 'factory farming'. Even larger industrial firms entered the industry. This was in the late 1970s when takeovers were the order of the day. Thus Imperial Tobacco Ltd., who owned Ross Poultry Ltd., purchased J.B. Eastwood Ltd., merging parts of both to form separate specialised divisions and later selling the whole poultry group to Hillsdown Holdings plc.

The men and women who were at the forefront of the emerging industry had entrepreneurial skills of a high order. There is no doubt that without their foresight and their ability to use scientific and technical information from many different sources, and to manage extremely large and complex enterprises, progress in adaptation of the American broiler industry to British conditions would have been long delayed.

Antibiotics as feed additives

In the United States the usual diets for chickens contained some protein derived from animal sources. When attempts were made to economise by using the cheaper plant sources, it was found that growth and the hatchability of the eggs were reduced. An animal protein factor was postulated; further work showed that it was probably Vitamin B12 plus other vitamin-like factors. A potent source of animal protein factor was a residue from the fermentation industry producing the antibiotic aureomycin. In 1949 F.L.R. Stokstad and T.H. Jukes of the American Cyanamid Company showed that it resided in the antibiotic itself. A number of antibiotics when fed continuously at subtherapeutic levels had the effect of increasing the rate of growth of pigs by about 15% and of chicks by up to 10%, with consequent more efficient use of their food. This was a very considerable economic gain and the finding was adopted by intensive pig and poultry farmers throughout the world, including the United Kingdom.

The reason for the response might seem obvious: the antibiotics controlled some low-grade infection within the animals. This was, however, by no means certain; other mechanisms were postulated. Marie Coates at the National Institute for Research in Dairying, Reading, showed that germ-free chicks, free of all microorganisms, did not respond to antibiotics, whereas outside the special incubators similar birds did. She later went on to isolate the particular organism that was specifically inhibited by the antibiotic.

Monensin, a narrow-spectrum Gram-positive antibiotic, which is used for treatment of coccidiosis in chicks (Chapter 11), was found to be effective as a growth promoter in cattle and was approved in 1976 as the first rumen additive for cattle which would improve feed efficiency. Doubt remains about the details of its mode of action but it has been clearly demonstrated that it improves feed efficiency by reducing feed intake (Muir, Stutz and Smith, 1977).

The addition of antibiotics to feed for pigs, poultry and calves in the United Kingdom became widespread. At the same time the public health authorities became concerned about the increasing incidence of resistance to antibiotics in organisms causing human disease, particularly the *Enterobacteriaceae*, a group that includes the salmonellas. Some of the resistance undoubtedly stemmed from the widespread medical use of antibiotics, and some from their therapeutic use in veterinary medicine, but a component could come from their use as growth promoters in farm livestock. A government committee was set up in 1968 under the Chairmanship of Professor M.M. Swann (Lord Swann), Vice-Chancellor of the University of Edinburgh at the time. The Committee recommended a division of antibiotics into two groups, those reserved for therapeutic use and those that were permitted for use as feed additives. The latter have continued to be effective and the United Kingdom legislation –

which was the first of its kind in the world – furnished a model that many other countries (except the United States) were to follow.

In contrast, the use of probiotics, which are substances and organisms thought to contribute to intestinal microbial balance (cf. the use of *Lactobacillus bulgaricus* in yoghurt by humans) and improve performance, carry with them no concern about adverse effects. The difficulty here is to prove the beneficial effects (Sissons, 1989).

The early weaning of piglets

The sow has a pregnancy of four calendar months. In the 1930s it was usual for the piglets to suckle her for eight weeks; when weaned, the piglets were fully capable of consuming solid food. The sow does not come into season to take the boar during her lactation but does so about a week after the piglets are weaned. If she fails to become pregnant then a complete oestrus cycle of three weeks has to elapse before she can be served again. With the weaning practices of the 1930s the total time for a reproductive cycle from the birth of one litter to that of the next was always in excess of 6 months. To maximise the number of baby pigs born in a year, one way would be to reduce the time for which the sow suckled her offspring.

Much was known through nutritional studies, notably by I.A.M. Lucas and G.A. Lodge, who had weaned baby pigs experimentally at early ages, about the nutrients that piglets required. This information allowed the formulation of suitable substitutes for sow milk. It was believed, however, that artificial milk could be given only after the piglets had been fed the natural first milk of the sow (the colostrum) since it contained antibodies to protect the piglet against infection and also promoted a process of gut closure which prevented absorption of foreign molecules including infectious agents. A series of investigations followed, notably at the National Institute for Research in Dairying by R. Braude in 1953 (Braude and Newport, 1977). He weaned piglets at two days of age. A system was also developed in Germany by H. Biehl in which weaning was at four days and one by Beecham Agricultural Products which was based on a solid substitute for sow milk. Later, in the early 1970s, A.S. Jones and V. Fowler at the Rowett Research Institute studied weaning at birth, without colostrum, as part of a programme to re-establish the Institute's pig herd by removing the gravid uterus from sows at full term and delivering the offspring into a sterile environment (see Chapter 11). In all these trials results were variable and difficulties ensued.

The first problem was that although the sow exhibited behavioural oestrus within a few days of weaning, the oestrus was not usually associated with ovulation: an oestrus with ovulation was more certain at about 18 days after birth. There was evidence, too, that matings after early weaning were not as fertile as matings under the former system of eight-week weaning. The losses in this way scarcely made up for the gains through shortening the weaning period. The second problem related to infection of baby pigs – with high variability from litter to litter – often resulting in a high mortality. The complex immunological problems of infection of baby pigs were not solved; although Braude's system was operated satisfactorily under what may be called laboratory conditions, it was not viable under farm conditions. The number of farmers who tried the new technology was very small.

It was really the farmers who made the practical advances; they appreciated fully that there was advantage in reducing the time the piglet suckled and in increasing the number of pigs a sow could produce in a year. Helped by the knowledge derived from the laboratory studies they reduced the weaning time first to six weeks and then, as they gained in confidence, to three weeks. The weaning age has now stabilised at three weeks throughout the U.K. This makes it possible to produce about two and a half litters per year. Comparing 1936 figures with those of 1986, the annual number of pigs reared per sow has risen from about 15 to about 23 and some herds have achieved 27 piglets per annum.

Non-nutritive growth promoters

The role of copper and antibiotics in promoting growth has already been mentioned. At the same time evidence was accumulating for the importance of other materials in promoting growth. The first to be followed up were the sex hormones and their synthetic equivalents, which soon came to be widely used by farmers, although more recently they have been banned because of possible dangers to the health of the consumer.

Sir Charles Dodd, of the Courtauld Institute of Biochemistry, had become intrigued by the diversity of the chemical structures of the naturally occurring oestrogens, and commenced work on the relationship between structure and activity. He was particularly interested in the high oestrogenic activity of the alkyl derivatives of a phenol known as anol, and in 1938 isolated diethylstilboestrol, an alkyl-substituted dihydroxystilbene. A family of similar substances was synthesised, including hexoestrol and dienoestrol; collectively they were known as the stilbene oestrogens.

Diethylstilboestrol was first used in animal husbandry by S.J. Folley during the Second World War to induce lactation in heifers that failed to breed. Folley developed a technique for implanting oestrogens beneath the skin and measured rates of release of the drug. This work was never applied. The major advance came in 1954 when F.N. Andrews, W.M. Beeson and F.D. Johnson in the United States found that diethylstilboestrol increased growth rate in steers and increased their conversion of feed to body weight. Much work followed in many countries with ample confirmation of these findings; because the stilbene oestrogens were cheap and some could be given by mouth they were used widely, particularly in veal production. The drugs were also used on human patients in the United States. In the early 1970s a number of cases of adenocarcinoma appeared in young pregnant women who had been given diethylstilboestrol. As a result the drug was banned by most European countries and by the United Kingdom, and the stilbene oestrogens are now not used in animal husbandry.

Considerable work then took place by the pharmaceutical companies to produce non-stilbene anabolic agents which included the natural sex steroids (notably oestradiol 17-beta), the synthetic androgen trienbolone acetate and an oestrogen zeranol, which was a chemical derivative of zearalenone, obtained from a fungus infecting maize (Chapter 11). These were marketed under trade names; although much more expensive than diethylstilboestrol, they were widely used by farmers in the late 1970s and 1980s with excellent results, particularly in castrated beef cattle.

The meat-producing industries of the United Kingdom have for generations been based on the use of castrated male cattle, sheep and pigs, together with surplus females. The reason, in the past, has been the danger of caring for intact males, which can be aggressive towards people in unpredictable ways. A further problem relates to the presence of an undesirable taint in the meat and bacon from male pigs. This was shown by R.L.S. Patterson of the Meat Research Institute to be due to a particular compound related to the androsterones; considerable work has been done to discover whether there are dietary or other ways to eliminate it. Despite this emphasis on the castrated animal it had long been known that, in all species, the growth rate of the entire male exceeds that of the castrate and also contains more lean tissue and less fat. Thus a bull grows about 10% faster and has about 5% more lean than a castrate; a ram grows about 6% faster and has 10% more lean in its carcase than does a castrated wether; with pigs, the boar has an advantage in both respects of about 2%. It can be wondered whether the effects of the anabolic agents might simply be replacement therapy for the missing testes, but effects of the anabolic agents appear to be in excess of the effects of simple replacement. Thus two large-scale trials (with over 2000 animals) conducted by the Meat and Livestock Commission and the Agricultural Development and Advisory Service showed that increases in live weight gain of 33% and 26%, respectively, were obtained by giving both androgens and oestrogens simultaneously. The use of anabolic agents was taken up widely by beef farmers.

With pigs, responses were found to be much smaller than in cattle, and the use of the new compounds by pig farmers was small. There was no great adoption of these methods by sheep farmers.

As with the stilbene oestrogens, concern was expressed that the compounds might constitute a hazard to human health. This concern was recognised during the 1970s and early 1980s but it was not until the 1990s that the compounds were banned under E.E.C. regulations. Whether the ban was justified is still a matter of dispute. The complex politics of the matter can be seen in the work of the committee set up by the European Community to report on the safety of these substances, under the Chairmanship of Professor G.E. Lamming. The work of the committee was stopped by the E.E.C. but the members of the committee felt impelled to publish their findings on the safety of trenbolone and zeranol (below) separately in the *Veterinary Record* (Lamming, 1987).

Several other types of compounds have been examined for their effects on the production of animals, including monensin, referred to above and in Chapter 11. Other substances that affect growth are the beta agonists, chemical compounds similar to noradrenaline. Noradrenaline has the properties both of a neurotransmitter of the sympathetic nervous system and of a hormone; some of its chemical relations (e.g. climaterol) have effects on growth rate of cattle steers. The effect appears to be one of repartitioning energy surplus to maintenance, towards protein deposition at the expense of fat (Hanrahan, 1987).

The somatotrophins are pituitary growth hormones which have been found to affect markedly the growth and efficiency of feed conversion in animals. The pituitary growth hormone of each species is unique; although their effects are similar in all species, during the period of this history they have found use in a number of different ways. Clearly, the use of natural pituitary extracts on any scale is impossible. It has been found that the material can be produced using recombinant DNA technology whereby modified bacteria (*Escherichia coli*) can be programmed to produce the material in useable quantities. The porcine somatotrophin (PST) has been used to promote growth in fattening pigs; that from cattle, bovine somatotrophin (BST), has been used to increase milk production by cows. In pigs the weight gain by the use of these materials is significant but the possibility of adverse effects on, for example, the mobility of the pig has not yet been ruled out. In cattle the production of milk is significantly increased and the difference in cow biology between BST-treated cows and untreated cows has been likened to the difference between cows of high genetic potential and those of lower potential. Great caution is being shown in licensing these materials for commercial use in all species. The economics of their use will also require to be evaluated. But above all these technologies raise ethical problems. They are probably not cruel; they can probably be brought to a technical level where the health of the animal or the health of the consumer is not at risk; but they raise the very large question of how far the human race wishes to go in manipulating its domestic animals for its own profit.

Selected references and further reading

Agricultural Research Council (1967). *The Nutrient Requirements of Farm Livestock*. No. 3. *Pigs*. Commonwealth Agricultural Bureaux, London.

Agricultural Research Council (1980). *The Nutrient Requirements of Ruminant Livestock*. Commonwealth Agricultural Bureaux, Farnham Royal, U.K.

Armstrong, D.G. (1981). Feed additives with particular reference to growth promotion. In *Recent Advances in Animal Nutrition in Australia*, ed. D.J. Farrell, pp. 146–59. University of New England, Armidale, N.S.W., Australia.

Blaxter, K.L. (1989). *Energy Metabolism in Animals and Man*. Cambridge University Press.

Braude, R. and Newport, M.J. (1977). A note on a comparison of two systems for rearing pigs weaned at two days of age, involving either a liquid or a pelleted diet. *Animal Production* **24**, 271–4.

Buttery, P.J., Haynes, N.B. and Lindsay, D.B. (eds) (1986). *Control and Manipulation of Animal Growth*. Butterworths, London.

Cole, D.J.A. (1978). Amino acid nutrition of the pig. In *Recent Advances in Animal Nutrition*, ed. W. Haresign, H. Swann and D. Lewis, pp. 59–72. Butterworths, London.

Coates, M.E., Fuller, R., Garrison, G.F., Levy, M. and Suffolk, S.F. (1963). A comparison of the growth of chicks in the Gustaffsson germ free apparatus and in a conventional environment with and without dietary supplements of penicillin. *British Journal of Nutrition* **17**, 141–50.

Crew, F.A.E. (1967). Reginald Crundell Punnett. *Biographical Memoirs of Fellows of the Royal Society* **13**, 309–26.

Dickens, F. (1975). Edward Charles Dodds. *Biographical Memoirs of Fellows of the Royal Society* **21**, 227–67.

Franklin, K.J. (1953). *Joseph Barcroft 1872-1947*. Blackwell Scientific Publications, Oxford.

Galbraith, H. and Topps, J.H. (1981). The effect of hormones on the growth and body composition of animals. *Nutritional Abstracts and Reviews*, series B **51**, 521–41.

Hanrahan, J.P. (ed.). (1987). *Beta-agonists and their Effects on Animal Growth and Carcass Quality*. Elsevier Applied Science, London.

Heap, R.B., Prosser, C.G. and Lamming, G.E. (eds) (1989). *Biotechnology and Growth Regulation*. Butterworths, London.

Hill Farming Research Organisation (1979). *Science and Hill Farming*. H.F.R.O., Edinburgh.

Hobson, P.N. (1988). *The Rumen Microbial Ecosystem*. Elsevier Applied Science, London.

Jones, A.S., Fowler, V.R. and Yeats, J.C.R. (1972). *The Improvement of Sow Productivity*. The Rowett Research Institute, Aberdeen.

Lamming, G.E. and 16 European scientists (1987). Scientific report on anabolic agents in animal production. *Veterinary Record* **121**, 389–92.

Leaver, J.D. (1988). Intensive grazing of dairy cows. *Journal of the Royal Agricultural Society of England* **149**, 176–83.

McDonald, P., Edwards, R.A. and Greenhalgh, J.F.D. (1981). *Animal Nutrition*. 3rd

Edition. Longman, London and New York.

McMeekan, C.P. (1956). Grazing management and animal production. *Proceedings of the 7th International Grassland Congress* (Massey Agricultural College, Palmerston North, New Zealand, 6-15 November 1956), pp. 146–56. International Grassland Congress, Box 1500, Wellington, New Zealand.

Mott, G.O. (1960). Grazing pressure and the measurement of pasture production. *Proceedings of the 8th International Grassland Congress* (University of Reading, UK, 11-21 July 1960), pp. 606–11. International Grassland Congress, Reading, UK.

Muir, L.A., Stutz, M.W. and Smith, G.E. (1977). Feed additives. In *Livestock Feeds and Feeding*, ed. D.C. Church, pp. 27–37. D.C. Church, D. & B. Books, 1215 N.W. Kline Place, Corvallis, Oregon 97330, U.S.A.

Sissons, J.W. (1989). Potential of probiotic organisms to prevent diarrhoea and promote digestion in farm animals – a review. *Journal of the Science of Food and Agriculture* **49**, 1–13.

Swann, M. (Lord Swann) (1969). *Report on the Use of Antibiotics in Animal Husbandry and Veterinary Medicine*. Command 4190. Her Majesty's Stationery Office, London.

Telford, M.E., Holroyd, P.H. and Wells, R.G. (1986). *History of the National Institute of Poultry Husbandry*. Harper Adams Agricultural College, Newport, Shropshire.

PART THREE
How did the science-based revolution happen, and what is the way forward as support is withdrawn?

Agriculture shares many of the characteristics and problems of manufacturing industry but it is peculiar in a number of ways. It is, and has always been, particularly vulnerable to changes in the local weather. Production is seriously curtailed in weather that differs from the mean significantly by being very cold, very wet and very hot and dry. These responses to the environment may be made worse by plant diseases and animal diseases, themselves sometimes responsive to climatic change. In the past the result has been famine and high prices. By contrast, especially favourable conditions can enhance plant yields and animal production and produce a surfeit of food with resulting glut and the collapse of prices. In the preceding section we have demonstrated how science has provided the farmer with a greater degree of control over these variables affecting his business so that year on year production of both plants and animals is more stable. This has allowed better planning and budgeting for the farming business.

A second characteristic of agriculture is that, because most countries are at least partly self-sufficient in staples, changes (even marginal ones) in the level of production from year to year result in great fluctuations in the price of world commodities and perturbation in the farmers' annual income which is largely outwith their control, with consequent severe social disruption.

In an effort to moderate these effects, which otherwise threaten to extinguish agricultural industries, governments have had recourse to various forms of protection and price support; the policies of British governments in our fifty-year period are outlined in Chapter 2. But as well as fiscal support British governments have provided support for agricultural scientific research and development, agricultural advice and agricultural education as important aids to the farming industry. The belief is that an industry in the forefront technologically will compete better in an economically turbulent world.

This section outlines the history of support for agricultural research, development, advice and education. It assumes that the technological advances reported in part 2 justify the support already given and asks what will happen in the future. Historical changes, changes in world trading patterns with world-wide free trade sought, and an absence of political commitment to support for the industry, all mean that support for agricultural research, development, advice and education is likely to grow less. Indeed, there is clear evidence that it is already less. In such a world and a world currently producing enough to supply all those who can pay for their food, the question arises: can British agriculture survive, and what part might government-supported science and development play in its survival? The need for clear technological objectives and what Blaxter calls technological optimism in framing the science policy which might mitigate the position of the agricultural industry is stressed.

13 Science during the revolution

Despite the fact that agriculture is closely woven into our history, lives and consciousness, and is held in warm affection by many, it is just another industry subject to the same economic laws as the pin-making factory of Adam Smith's imagination. The extra dimensions in the agricultural industry are: its exposure to the fluctuating competition of the world market place; the acute and unpredictable impact of the variations in the weather; and its perishable products, which can be stored only with difficulty and at considerable cost. Over the 50 years from 1936 to 1986, there has been an attempt to mitigate these impediments through the provision by successive British governments of a research service and cost-free advice and support as well as the substantial price and market support provided during most of that period (Chapters 1 and 2).

Government support for science

Interest in this chapter is focused on the scientific support for agriculture. As already related (Chapter 2) the earliest allocation of funds for agricultural education was the 'whisky money' of 1896, which, thanks to the efforts of Mr H.D. Acland M.P., was made available to county councils for rate relief and technical education. For many U.K. counties, technical education meant agricultural education; many agricultural colleges were begun or extended at this time. Their interest for us is that the lecturing staff began to be used as advisers and often carried out simple field experimentation to solve problems of manuring, varietal value, etc.

Another important piece of legislation was the setting up of the Development Commission under the Development and Road Improvements Fund Act of 1909. Its function was to advise the Treasury on the making of grants or loans for the development of rural and coastal areas, including grants for agricultural research. Although its means were slight it continued to grow in strength and we find it placing money strategically with nascent research institutes and university departments. It had far-seeing men like Sir Thomas Middleton and Sir Daniel Hall on its Board.

The next formal reference to 'research' by Government was in 1921 with the hasty repeal of the Corn Production Act of 1917 and the substitution of a small acreage payment totalling £19 400 000, with an additional £1 000 000 set aside for agricultural education and research. This allocation was made with the aim of helping the farmer in a difficult situation because of lack of protection from foreign imports. A belief that science could somehow improve the British farmer's competitive position in a free market has persisted down the years and has been the main policy justification for the Agricultural Research Service.

The next policy advance came as the National Government in 1931 sought to redefine policy for the agricultural sector. There were four thrusts to this policy: marketing reorganisation (the origin of the Marketing Boards); regulation of imports (quotas and import duties); subsidies and price insurance; and measures to increase efficiency, including subsidies for draining and lime, licensing of approved sires, the provision of a national veterinary service, and statutory schemes for the eradication of animal and plant disease. Added to these was a provision for agricultural education and research.

Murray estimates an annual disbursement for agricultural research of about £3 000 000 from all sources, including the Development Commission, in the years immediately prior to the Second World War. Between the wars the Development Commission became the largest single source of government support for agricultural research. Lesser amounts were disbursed by the Board of Agriculture, which later became the Ministry of Agriculture, and by the Scottish Department (formerly Board) of Agriculture and the Department of Agriculture for Northern Ireland.

If there was a realisation that scientific research might help agriculture and if funds were becoming available it was still by no means clear how they were going to be applied for the benefit of the industry, nor what a suitable departmental structure might be for their administration.

The early Research Institutes

Until this time funds had been applied *ad hoc* wherever it was thought that they might be used to best advantage. The great centre for agricultural research was Rothamsted Experiment Station (1843) and it naturally became the role model for later events. In its early years Rothamsted had a clear goal in the investigation of soil fertility. It was controlled by the Lawes Trust, derived from the sale by Sir John Lawes of his fertiliser business. As a result for many years it was independent of government funding and indeed even today, when its activities have moved ahead of that level of funding, it still has a measure of independence through the private administration of the Trust Funds. A similar initiative came from the cider growers of the West Country, who set up the Long Ashton Research Station near Bristol (1903), a collaborative effort initiated by Robert Granville with help from the Bath and West and Southern Counties Agricultural Society, and subventions from the county councils of the area. Later it became closely associated with the University of Bristol. The benefaction of one man, John Innes, set up the Horticultural Institute (1910), which bears his name, and maintained it fully until just after the Second World War, when its remit widened and its activity increased. The Rowett Research Institute (1912) was set up with the help of the Development Commission and the Scottish Education Department but benefited from the specific benefaction of J.Q. Rowett and a number of local benefactors. Then came the East Malling Research Institute (1913), the Welsh Plant Breeding Institute (1919), the Animal Diseases Research Association at Moredun (1921), the Scottish Plant Breeding Station at Pentlandfield (1921), the Hannah Dairy Research Institute (1928) at Auchencruive, and the Macaulay Institute for Soil Research at Aberdeen (1930). All these were privately funded at the start, either by single individuals or by groups of concerned farmers, and their funds later topped up by Development Commission funding or funding from the appropriate Boards, Departments or Ministries. Two other research institutes of that time, the National Institute for Research in Dairying at Reading (1912) and the Plant Breeding Institute at Cambridge (1912), were founded within their respective universities but with funding from the Development Commission and the Board of Agriculture, although the initiative no doubt came from within the universities. The Animal Virus Research Institute, which began as the Pirbright Experimental Station (1924), was set up by the Ministry of Agriculture with the special purpose, originally, of carrying out experimental work on foot and mouth disease (Chapter 11).

As each of the research institutes was formed it had an immediate goal, and the policy for its achievement could be managed easily by a Board of Governors and a Managing Committee or other administrative mechanism, with the day-to-day involvement of the director and, generally, a very small staff. Moreover, their areas of work scarcely overlapped. Once the immediate objectives had been reached these institutes began to look for further work and an able director would have expansive ideas. A realisation that there was an increasing number of these research institutes, as well as university departments and colleges, engaged in research for agriculture and in part publicly funded, often from a variety of sources, and with no single oversight of their programmes, rang alarm bells in government.

The Research Councils

As a result of the deliberations of the Haldane Committee, set up by Lloyd George in 1917 to study the machinery of government, the Medical Research Council (M.R.C.) was instituted in 1918 under a Committee of the Privy Council so as to be free of the routine demands of its cognate Ministry of Health. In this it was following a pattern already set by the Department of Scientific and Industrial Research (D.S.I.R.) in 1916, and the same pattern was followed in the setting up of the Agricultural Research Council (A.R.C.) in 1931. These quasi-independent bodies were looked on with suspicion by the administrative departments of government and there were several attempts to bring them within the normal administrative framework. One of these sensitive departments was the Ministry of Agriculture. In fact the Haldane Committee had envisaged a separate Ministry of Science with responsibility for the M.R.C. and other such bodies in the future. In course of time this happened and the Research Councils (now including also the Natural Environment Research Council, the Science and Engineering Research Council [former D.S.I.R.] and the Social Sciences Research Council) came under the care of the Department of Education and Science, thus retaining their independence within a department with a more general remit.

The A.R.C. was set up in 1931 under the Chairmanship of Lord Richard Cavendish. Its first Secretary was Sir William Dampier, who was then Secretary to the Development Commission. The details of the early Council and its development are admirably laid out by Sir William Henderson (1981); suffice it to say that at first the role of the A.R.C. was very circumscribed, being largely advisory to the Agricultural Departments (the Ministry of Agriculture and the Department of Agriculture for Scotland) and the Development Commission. The control of the Research Institutes (then funded to the surprisingly small amount of £390 000 per annum) remained with the Agricultural Departments. To add to the confusion, the parliamentary responsibility for agricultural research was to stay with the departmental ministers. The A.R.C. was given £5000 to use on its own authority. This situation remained until 1935, when the administrative arrangements were changed to permit the A.R.C. to carry out research itself, but this was still a constrained situation. Some amelioration took place as a result of the deliberations of the Hankey Committee (1941) when the A.R.C. was placed on the same footing as the M.R.C. and D.S.I.R. The situation was further corrected by the passing of the Agriculture Research Act in 1956, which placed the financial control of the Agricultural Research Institutes in England and Wales with the A.R.C. although for reasons of national pride the Institutes in Scotland remained the responsibility of the Department of Agriculture for Scotland (D.A.F.S.), with the very close involvement of the A.R.C., which advised the D.A.F.S. on scientific policy, using Visiting Groups and (at one time) scientific advisers (see below). Staff gradings and promotions procedures were common to both. That is why the collection of Research Institutes and Research Departments in England, Wales, Scotland and Northern Ireland is referred to collectively as the Research Service. One other recommendation of the Hankey Committee, that measures should be taken to attract workers of outstanding ability into agricultural research, was an important one and bears on the success and failure of the system.

Policy for Research Institutes

With some 32 research institutes (eight in Scotland) in place in 1931, a number which scarcely altered over the next 50 years although the mix of subjects did, the Agricultural Research Council had two problems to tackle. How did it control these organisations, and how did it ensure that initiatives at the cutting edges of science were supported without multiplying or over-inflating its institutes? The first problem was met by delegating powers to the governing bodies and directors for the detailed policy of the institutes. Coordination of programmes was achieved by a number of in-house (A.R.C.) scientific advisers, who visited the institutes to discuss specific aspects of programmes and to ensure by discussion and by relaying information that excessive duplication between institutes was avoided. To monitor the standards of performance of the institutes, review groups (Visiting Groups) were set up, which visited the individual institutes perhaps every four or five years. The members of the group were scientists, whose ability would command respect within the institute, as well as eminent agriculturalists and A.R.C. Headquarters staff. In the early days the remit for these Visiting Groups was broad. In later years their remit became curtailed, particularly in regard to the organisation of the scientific programme, under pressure from other programme-designing procedures.

New scientific initiatives

The second problem was the encouragement of entirely new scientific initiatives. The Council sought to achieve this by the creation of units, mostly within University Departments or occasionally Research Institutes, to support research workers who were judged by their peers, in appropriate standing committees, to have identified an important and fertile new field of research and to have the necessary qualities to lead work in it. These units were largely *ad hominem* and disappeared when their directors retired. Sometimes they grew and became either institutes in their own right or parts of institutes. Later, in the 1970s, as funds became restricted, important areas of work thought to be required for the improvement of the industry were set up between groups in universities and/or research institutes to concentrate on a particular aspect, e.g. increasing the efficiency of photosynthesis. These research groups, which were already part-funded by their own institution, could be further funded much more cheaply.

The tightening purse strings

These were halcyon days. The first clouds began to appear on the horizon with the appointment by Harold Macmillan in 1962 of a Committee of

Enquiry into the Organisation of Civil Science, under the chairmanship of Sir Burke Trend (later Lord Trend). The reason for this was the burgeoning expenditure on civil science and the realisation that, with developments in the pipeline, costs of equipment and staff were likely to continue to expand. The report of this committee was followed by the Science and Technology Act of 1964, which incorporated some of its recommendations, including the designation of the Department of Education and Science as the Department answerable for civil science in the House of Commons. The act also set up the Council for Scientific Policy. None of these troubled the Agricultural Research Council unduly. Other suggestions about the role of the Secretaries of the Research Councils did; they were eventually withdrawn. Further suggestions on the range and mix of responsibilities of the Councils caused initial disquiet but eventually were accepted with some minor modification. Eventually the Council for Scientific Policy, on which the Research Council Secretaries were assessors, was transformed into the Advisory Board for the Research Councils, on which they had full membership.

The next clouds were less visible and harder to report dispassionately. The perceptions of those who administered research and development organisations during this period are perhaps not wholly objective but may be worth including. Perhaps only 'The toad beneath the harrow knows exactly where each tooth-point goes' (Kipling).

The first change was the growing awareness of the need to achieve value for money spent; the second was the appreciation within government ministries of the need for 'accountability' such that, at every level of the civil service, individuals felt they had to be prepared to account for expenditure and in so doing had to have more and yet more information from their funded organisations to enable them to justify the money spent.

Rothschild and Dainton

The Agricultural Research Council in 1966 set up a Working Party to advise the Council on the direction and depth of future research. It produced some excellent general guidelines for the principles governing the selection of future programmes. But broad guidelines were not enough and in 1971 a Green Paper (Cmnd. 4814) was produced under the authorship of Lord Rothschild on 'The organisation and management of Government R & D', and including, under the authorship of Sir Frederick Dainton, a Report on the 'Future of the Research Council System', produced by a Working Party of the Council for Scientific Policy.

No-one can gainsay the value of these papers. The Dainton Committee Report, in part because of its remit, comes up with limited changes to the system then in force but it is the result of detailed analysis and provides very valuable information on which to make judgements of the system. Thus it shows that for the year 1971–72 the share of the Science Vote Estimates for the Research Councils was: A.R.C., 16%; M.R.C., 19.6%; N.E.R.C., 13.6%; S.R.C., 47.5%. Moreover, the average annual percentage increase in real terms for the science votes over the period 1965–71 was: A.R.C., 6.7%; M.R.C., 8.4%; N.E.R.C., 14.6%; S.R.C., 6.2%. The value of the A.R.C. vote in 1971 was £18 740 000 out of a total of £117 247 000.

The Rothschild report, which was published in the following year as a White Paper, was much more revolutionary and perturbing. It identified the need for a customer–contractor relationship between (in this instance) the Ministry of Agriculture and the A.R.C. In this relationship the customer defined the research required to be carried out, and paid for it. The contractor undertook to deliver and received appropriate payment. This admirable concept had a number of consequences. First it was a bureaucrat's delight within the civil service and multiplied the paperwork in research organisations, which were not staffed or organised to provide it. Within a research institute, it distorted the director's capacity to determine the programme and to balance it with the abilities of staff. Managing scientific staff of high quality, with the eccentricities and independence of spirit which they often demonstrate, is not helped by having outside organisations also shaping the programme. There were other general consequences, too, which will be considered later.

Once the Rothschild proposals were made firm in the White Paper 'Framework for Government Research and Development' (Cmnd. 5046) in July 1972, a series of modifications of the previous procedures for deciding agricultural research policy and funding emerged. There was first a transfer of funds from the A.R.C. budget to that of the Ministry of Agriculture so that by the third year of the transfer 54% of the Council's share of the science budget would have been transferred to the Ministry. Then the customer departments had to set up a Chief Scientists' Organisation to decide what research they wished to fund. The A.R.C.'s Charter had to be amended to include members of the customer departments. To facilitate dialogue between the customer and contractor and to ensure a useful input of the views of the industry, a Joint Consultative Organisation was set up, consisting of an Animals Board chaired by a farmer, W.A. Biggar, an Arable Crops and Forage Board chaired by another farmer, J.S. Martin, a Horticulture Board chaired by a nurseryman, D.G. Frampton, an Engineering Board chaired successively by two academic engineers, Professor A.R.J.P. Ubbelohde and Sir Hugh Ford, and a Food Science and Technology Board chaired by another academic, Professor A.G. Ward. Each Board was served by a comprehensive series of commodity or subject committees. In Scotland the interface between the Department of Agriculture and its contractors, research institutes and agricultural colleges, was overseen by the Scottish Agricultural Development Council (1972) chaired by a farmer and businessman, Sir Michael Joughin.

The resulting bureaucracy

Never in the history of British agriculture has so much discussion taken place over the direction and nature of the agricultural research required. At the end of all the discussion, as Sir William Henderson says 'It was of some satisfaction for the Council to be able to note that each Board had substantially endorsed the existing programme of research in its subject area'. Moreover, he notes that the funding available had begun to grow less year by year so that many of the innovative areas suggested by the Boards could not be pursued, to the frustration of Board members. It was found impossible to sustain the impetus of the discussion on these Boards; they were first modified into a single Joint Consultative Board and finally closed down altogether in 1983. In hindsight this was a very big hammer to crack a very small nut, and a great many Board and Committee members may have wondered if they had spent their time profitably.

Nevertheless, the operation served the purpose of involving a large number of agriculturalists and scientists in the process of slimming down a research and development programme which had outgrown the funds available and which was entering a new phase with a government in power which was less sympathetic to the support of agriculture, apparently less concerned about its contribution to the balance of payments, and committed to the nation's prospering by non-agricultural commercial activity. The Joint Consultative Organisation and its Boards also left behind a large number of very useful reports, which formed a background for those in the A.R.C. and the Ministry of Agriculture concerned with the commissioning of research, which, under the new system, was an ongoing process. The Joint Consultative Organisation was replaced by the Priorities Board (1984), a much streamlined organisation, which advised on the broad allocation of public funds to research and development, helped by its Research Consultative Committees, which advised on the work required within each sector. The role of the Ministry of Agriculture gradually became stronger, with the involvement of its policy divisions in the commissioning process. This was useful in affording another route by which the needs of the industry could be expressed, but less useful in perhaps focusing on shorter-term projects.

The A.R.C. had another very important role, that of promoting research in university and college departments. After the Rothschild changes this remained as a function that it was allowed to carry out without the help of outside agencies. It did so through its Research Grant Board, which functioned in two parts, the Plant and Soil Board and the Animals Board, where proposals for new areas of work from academic scientists were judged by their peers and funded modestly, often with very valuable results. It should be noted that with the advent of the E.E.C. another element entered the funding of agricultural research with funds becoming available through a complex system of bidding and evaluating in the E.E.C. bureaucracy, with the disadvantaged areas of Scotland and Wales now competing with similar areas in Portugal and Spain and adding a complicated and sometimes dispiriting gamble to the research planner's activity.

We have concentrated in the foregoing paragraphs on the changes in the Agricultural Research Service because that was the main agency for scientific innovation in agriculture in the United Kingdom at that time. But it was not alone.

Advice and education

When agricultural technical education was begun in Britain in the 1890s, an association grew up between the teaching of agriculture and the provision of specialist advice. The history of this association in Scotland is outlined by Fleming and Robertson (1990), and by McCann (1989) and Winnifrith (1962) for England. What is clear is that by 1936 the main agricultural colleges, and some university departments, in England had associations with the advisers, specialist advisers and departments of, for example, soil chemistry, which provided services to the farming community. At the end of the Second World War the National Agricultural Advisory Service was set up with Sir James Scott Watson as its first director. In the beginning it used staff who had been trained for specific technical promotional projects, such as the improvement of silage, as general advisers. Its specialist advisory services, such as soil chemistry and plant protection, were provided from its university- and college-based departments until it withdrew its advisory activities from them and set up its own regional headquarters, often nearby, and with local offices strategically placed. In Scotland the advisory service remained attached to the Colleges of Agriculture and continued so even after the 'privatisation' of the service. Preceding the White Paper 'Framework for Government Research and Development' was one entitled 'Proposed Changes in the Work of the Ministry of Agriculture, Fisheries and Food' (1971, Cmnd. 4564) indicating changes in the National Agricultural Advisory Service, which became the Agricultural Development and Advisory Service (A.D.A.S.) in March 1971, streamlined and reorganised with a saving of some £15 000 000 a year and a reduction in staff. The agricultural aims of the new organisation are set out very clearly and are worth quoting in part.

> Within the general aim of the Government the objective for home agriculture, having regard to the proper use of resources, is to provide the economic conditions required for the selective expansion of production from an efficient agricultural industry with a rising trend of productivity. This objective is pursued in a variety of ways. Apart from deficiency payments and production grants the Ministry provides certain other forms of direct support to encourage farmers to improve and invest further in

their enterprises; it provides advice on technical and managerial problems facing the industry; and it undertakes development work on experimental farms and horticultural stations to assess the practical application of the results of research and communicate them to farmers and growers. It administers schemes to promote animal and plant health and it undertakes applied research relating to these subjects and to pest control. It also undertakes world wide botanical research through Kew Gardens. It is concerned with many aspects of the utilisation and tenure of agricultural land ...

Support for agriculture

Until 1971 the Agricultural Development and Advisory Service had been available to farmers without cost and had been considered as part of the support system. From that time forward there was gradual introduction of charges for advice and eventually the whole advisory service was expected to stand on its own feet.

The A.R.C., which later within our period became the Agricultural and Food Research Council (A.F.R.C.), and the Agricultural Development and Advisory Service of the Ministry of Agriculture, Fisheries and Food (and their equivalent in Scotland and Northern Ireland) complement each other. It would be idle to enter into the semantic analysis that took place about this time seeking to demonstrate a continuum from basic science through applied science to mission-oriented science and development, in the innovation process. Fortunately all the analysis and the competition for funds between those engaged in science of various categories led to the appreciation of the complexity of the process of establishing new technologies and the all-embracing nature of development, which Dainton calls 'tactical research' and which he described as 'the science and its application and development needed by departments of state and industry to further their immediate executive or commercial functions'. He expands this further by saying that 'At one extreme it may contain a significant element of sophisticated research over a long period; whilst at the other extreme it amounts to little more than a modest intelligence and advisory activity'.

Nature of Development work

For many years a belief existed in a terminal activity called 'development', which prepared the discoveries of basic science for use in industry. It was believed that the science came first and the development followed, as it did in a very few instances. Indeed in the 1980s development agencies were asked to say what they considered ripe for development in the programmes of their cognate research institutes, and the institutes were required to offer appropriate material for development. Needless to say the process came to a dead stop because the time scale of research is not such that it is turning out new discoveries fit for development at a fast rate. A corollary to this thinking was the impression (and it can be no more objective) given by research administrators that research activity was intellectually more demanding than development activity. Perhaps the whole promotion and reward system of science was coloured by this belief. These would be trivial matters if they did not perhaps affect profoundly the rate of innovation in agriculture.

In fact development begins with the recognition of an industrial (or other) need and the deployment of an appropriate skill to satisfy this need. Sometimes it will be the need for advertising to establish a market, or the need for standardisation to secure that market; sometimes the recognition of an impediment to acceptance of the product (for example, the fatty acid mix in oil seed rape) and research designed to redress the imbalance; sometimes basic biochemical work is needed to reveal the nature of a toxic material (e.g. glucosinolates in oil seed rape) and genetics and plant breeding to modify the metabolism towards a less toxic state. The process does not stop there: work on other areas (in the case of oil seed rape, on methods of harvesting, feeding the meal and producing balanced rations) within an economic and management context is also required to make the process work. It was a recognition of this process that led Rothschild to propose his customer–contractor relationship. The problem in agriculture was that farmers were not fully equipped to articulate their needs to the research sector for action. They would often articulate the narrow needs of their own businesses but not the wider needs of their whole area of the industry. That was the reason for setting up the Joint Consultative Organisation. It was not a very satisfactory intermediary, controlling the input of Ministry (formerly Research Council) funds into Research Council research. It had difficulty in diverting funds to aspects outside the Research Council remit, such as marketing. Although mechanisms existed in the Ministry of Agriculture to take care of marketing, quality control, etc., it is not apparent how much innovative thinking could take place within a ministry designed, on the whole, for other purposes. The later involvement of the policy divisions of the Ministry of Agriculture in the commissioning of research had advantages in articulating the needs of the farmer very clearly but disadvantages in perhaps concentrating on more short-term research work.

The problems of innovation

The problems of innovation were examined in 1985 in a classic series of studies by the Agricultural Technology Group of the National Economic Development Organisation (N.E.D.O. or Neddy). The studies comprise five case studies of separate innovations (oil seed rape growing, hill sheep farming, silage making, energy screens for glasshouses, and modern cereal growing methods) and a general paper bringing together information derived from them. The first three of these are discussed here. They demonstrate the great complexity of the development process.

In the case of oil seed rape, a minor crop became important because of a change in E.E.C. support reflecting a change in policy and designed to establish a local oilseed industry, following doubts about the security of supplies of soya from the U.S.A. and other suppliers. Thereafter the development process followed. The first step was the transfer of technology from countries, such as Canada, which were already involved in the growth and export of the crop. Thereafter the advisory, development and research services worked closely together to produce locally suitable varieties and to incorporate genetic material with a reduced content of erucic acid, and eventually of glucosinolates, in the seed (see Chapter 8). The case studies also note the important role of merchants, in the supply of inputs, such as storage and transfer to crushers, as well as in encouraging the growing of the crop and providing technical advice to ensure success. In looking at this problem the committee noted the later development of the crop in Scotland and, while identifying a number of technical factors such as the slower arrival of winter barley in Scotland and the less pressing financial factors of rent and capitalisation (both less in Scotland), noted that the absence of a merchant network to help with the promotion was an important factor in the slow take-off. Likewise the image of oil seed rape as a southern crop was probably a further impediment to take-off.

In their review of improved grassland management in hill sheep farming systems, the N.E.D.O. committee are tackling a notoriously difficult and slow to change sector of agriculture. The story here begins with the establishment in 1954 of the Hill Farming Research Organisation (H.F.R.O.), an institute devoted to increasing production from hill farming (Cunningham, in Hill Farming Research Organisation, 1979). As such it was free to undertake the fundamental work required on, for example, the nutritional quality of heather, but also to undertake (with its development remit) an investigation of systems suitable for better production by the hill ewe. This was achieved by analytical research concerned with obtaining an understanding of the biological components of production. In a production system many components interact with each other. A full understanding of a system cannot therefore be obtained solely by studies of its components in isolation; research must be undertaken on systems. Systems research is concerned with putting components together and with evaluating the systems of production that result from such synthesis.

For the case study the N.E.D.O. technology group chose the 'two-pasture system' which was formulated by H.F.R.O. and developed and evaluated as the result of cooperative work between H.F.R.O., the Scottish Colleges of Agriculture and the A.D.A.S. Experimental Husbandry Hill Farms (Chapter 12). In this case commercial organisations played no part in the development because it presented little significant improvement in their market. The main effort had to be put into convincing farmers and shepherds that the system could be introduced to their farms without disrupting their normal annual cycle too drastically. It also involved adapting the system (which effectively dedicates an improved area of pasture for use at nutritionally critical biological periods such as tupping and post-lambing) to the circumstances of the individual farmer. Another element in this project was the careful financial monitoring of the individual farms, which was able to demonstrate a remarkable annual improvement in net farm income for all the farms monitored.

When the Group looked at silage they saw that the research work that disentangled the complexities of the silage fermentation process was the initial impetus to improvement in silage making. Its further development depended on the improvement in storage facilities, which were often helped by innovations by farmers, such as the Dorset wedge system (Chapter 8). Then it was necessary to improve machinery handling at harvest. The original flail forage harvesters did not chop and bruise the forage and easily contaminated the silage. They were followed by double-chop machines, which chopped the material reasonably finely, and eventually by precision-chop machines, which encouraged finer chopping and bruising and allowed an egress of sugar-bearing sap. Then methods had to be developed to utilise the silage, either by easy feeding methods or by adapting buildings and using side-delivery waggons. Other problems were the avoidance of effluent or its safe disposal, and improvements in the strength of side-walled clamp silos, which came under great pressure when fully filled. The Group also looked at the importance of additives and the development of an applicator to distribute them. Promotion of silage utilisation by the advisory services was helped by the interest of firms like I.C.I., who saw an opportunity for increased sales of the fertiliser required for improved grassland from which the silage was made (as yet there was no concern with the water-polluting effects of excessive fertiliser use). Once more, no one scientific step led to the development of a sound silage system. Many people, many disciplines and various degrees of research focus were involved in the establishment of the systems.

The N.E.D.O. Technical Group also studied reduced cultivations for cereals and thermal screens for glasshouses and came up with the same message: that the development process is a complex one with economic evaluation forming an important part of the process. Once more in this book tribute must be paid to the economics and farm management departments of the advisory services; often associated with University departments, they gathered the data and produced the methods of analysis which enabled sound advice to be given on business problems and also allowed objective data to be gathered for the evaluation of various innovations.

Support versus free trade

At the beginning of this chapter the question of the effectiveness of support to agriculture was raised, whether by price or market support or by scientific research. There is ample evidence that the support by deficiency payments, commodity boards and financial aid for improvements maintained a prosperous and stable agriculture and an insurance against world catastrophe, over the 50 years with which this book is concerned (Tracy, 1964). However 'You can't buck the market', at least not for very long. The deficiency payment system transferred income from taxpayers to the farmer, a perfectly legitimate procedure to achieve a particular policy. The deficiency payment is determined as the difference between what is agreed to be a reasonable price and the price of the commodity on the world market. If the deficiency payment trigger price is set slightly high it encourages production (another policy decision), so that gradually the bill for support becomes larger. Since, too, there is no restriction on imports, the prices of these can fall in glut years, increasing further the bill for support of home-produced commodities. When the bill for agricultural support became £343 000 000 in 1961–62 and remained at £321 000 000 in 1962–63, disquiet began to grow over the extent and the method of support; this continued to be called into question for the next ten years, although no radical change was made until Britain began to come under the influence of the E.E.C. But even as late as 1979 the White Paper 'Farming and the Nation' (Cmnd. 7458, 1979) was seeking to maintain and increase production using the E.E.C. mechanisms.

Inexorably from then on support came under pressure from the demand for free trade. At the time of writing the pressure to free world trade under the General Agreement on Tariffs and Trade is threatening to reduce agricultural support within the E.E.C. In this process those countries with a large number of small farms are most fearful of the change. Those where farm sizes are larger and with more highly developed technology can contemplate a number of difficult strategies, all of which are likely to leave them with a still smaller number of larger farms.

The amount and method of fiscal support for agriculture must always be a political decision in the light of local and world agricultural problems. However, fiscal support, probably of a temporary nature, can be a useful adjunct to science-based enterprise development, as we saw in the establishment of oil seed rape and of bread-baking quality wheat (Chapters 7 and 8).

Contributions of science to productivity

The contribution of science to farming productivity and to the well-being of nations can be judged from the small sample of improved production reported here. Attempts to quantify more exactly the value of scientific work for agriculture generally fail to take sufficient account of the interaction between government policy, market forces, and the individual circumstances of farmers seeking innovations within the subject area of the research and development work. A broad-brush approach can be more objective and, we hope, as meaningful.

Throughout the book there are numerous examples of elements essential for the growth of plants or animals which are absent from ancient rocks, leached sands or pumices, and from the soils derived from them. Such soils are almost totally useless for agriculture. Once the missing element is replaced, however, it allows a flourishing agriculture where none was possible before. Phosphorus deficiency in South Africa (Chapter 9) is a case in point: the provision of phosphorus-rich licks on pastures that otherwise failed to support healthy cattle released thousands of hectares of otherwise barren soil to productive farming. Similarly, locally dramatic cases of deficiencies in plants, such as boron and copper in peaches, zinc in cacao, sulphur in tea, manganese and copper in cereals, etc., prevent profitable cropping without the addition of the appropriate element. Less dramatic examples of soils that fail properly to support plant and animal growth, because of trace element deficiencies, are numerous in Britain. A crude assessment of value suggests that the research cost can be amortised over an infinite number of years, to the point of disappearance, so that there is an agricultural gain, offset somewhat by the reduction in price that extra production generally brings. That gain is only worthwhile if the newly fertile land produces animals or plants that can either provide exports or satisfy a demand within the country at a reasonable price.

The contributions of workers on plant and animal disease, pest and weed control can be looked at in the same way. Here are discoveries that allow crops or animals to flourish in situations where they could not flourish before. Many of these successes depend on the production of new therapeutic substances. Development of these substances has largely been a function of the private sector but the costs involved, the uncertainties of the rewards (because of the pressure of competing materials), the development of resistance in the pathogen, and the dangers of legal actions from unexplained toxicities or unexpected environmental effects, all suggest that in future there may be a case for the involvement of government research and development agencies to a greater degree.

The breeding of better (more productive or higher-quality) plants and animals also brings a gain. In animals, the breeding programmes have been able (in some species) to combine selection for faster growth rate with increased efficiency of food utilisation, so that if they are more expensive to keep this is balanced by the greater weight of saleable product. In plants, on the other hand, the gain in performance is not yet matched by a gain in efficiency, except in so far as the harvest index is improved in new cereals. Accordingly, increase in yield brings with it a requirement for more fertiliser, increased precision, increased machinery power and more skilful management. It becomes much more difficult to evaluate the contribution of high-yielding cereal varieties where the staple grains are not in short supply. This is not to undervalue the achievement but because there are so many ways of looking at it, in terms of profitability, land saving, fertiliser use, and so on. There would be no doubt about its value in a situation of general shortage, or where yield per hectare is an element in the farm budget. Most importantly, animal and plant breeding and the associated husbandry changes have brought about a greater stability from year to year and a greater predictability so that the farm can be handled like any other business with budgets and targets that are not upset by changeable seasons and variable food quality.

Other scientific advances mentioned in previous chapters, such as the increased understanding of soils, the development of measurement in experimental situations, and the deeper understanding of nutrition, each underpin work in crop husbandry, agronomy, plant and animal breeding and animal husbandry. They not only contribute advances that lead directly to improved farming methods but also provide information and knowledge to the farmer; these provide security and give the farmer greater control over his circumstances.

The areas discussed in the chapters on mechanisation and integrations in animal and crop husbandry are rather different. They are concerned with development, the evaluation of which is even more difficult.

Of mechanisation, it can be said that it has completely transformed modern farming and the life of the farmer and farm worker. It has brought flexibility and speed of working, and a reduction in labour and back-breaking tasks, although it has not removed completely the boredom and drudgery of spirit from field work. It has a downside, however, in that it brings high costs of energy to the system (Blaxter, 1977), high fixed costs to the farm business, and unemployment to part of the work force replaced. However, it is now a permanent part of the farming scene and, in spite of the attempts of some enthusiasts to reverse the trends and return to simpler methods of farming, overall productivity could not be maintained without it.

Integrations and innovations

The introduction of new farming methods show marked interactions with the farm business. For example, conversion of a farm to the use of silage is an expensive matter and hard business decisions have to be made before undertaking a change. Similarly, the profitability of increased sheep production by one of a number of new methods of managing pasture is in part dependent on the current degree of market or area support available, and on the ability of other countries to compete for the U.K. market.

The difficulty in evaluating development work of this kind is that it is aiming at a moving target. Because systems take time to organise and crops and animals take time to grow, the developer may be left behind by events such as shifts in markets or the success of alternative systems. The same is true of the basic work that may be called on to produce a solution to facilitate a particular development: it may be left behind by similar changes and frustrate the research workers, who gets no 'development' satisfaction and who may have eschewed some other more exciting scientific line of enquiry in its favour. That is one reason why it is difficult to engage the interest of good basic scientists in such work.

Nevertheless this development work, which is concerned with improving the efficiency of businesses, introducing new ideas and innovative practices and developing new systems, is the foundation of a flourishing industry, flexible enough to respond to new market challenges.

The roles of discovery and development

There are thus two elements in the science required for agricultural advance. There is basic work, focused to various degrees on problems identified within the industry; and there is work designed to reach a specific industrial objective, which we have referred to as development work, and which may include some sharply focused basic work or any one of a number of other activities (systems research, marketing, advertising, etc.). This type of work, which is well organised in other research-and-development-led industries, has not had as much attention as basic science in agricultural innovation. Where it has happened, it has been because of the remarkable network of people in the research and development system, farmers, advisors, developers and basic scientists devoted to agriculture. Their close association has allowed informal agreements, exchange of suggestions and pooling of resources to take place.

The 50-year period of the modern agricultural revolution has seen a rich reward from the exploitation of basic science, often by serendipity. The need for the next half century is for a more structured exploitation of science for the development of innovative business. This requires the informal network of today to be strengthened and shaped with the identification of multidisciplinary teams with designated leaders, sometimes from research, sometimes from development, but on an equal footing.

Aims for the industry have to be clearly identified against a stable policy background and project leaders, who must be given appropriate resources, identified to achieve them. Many of these resources are (or should be) in the hands of directors of Institutes or Experimental Husbandry Farms (recently many of the old institutes have been amalgamated and their directors have become Heads of Station within larger institutes). Directors (or Heads of Stations) must be free and trusted to divert resources to promising areas of work. This is not to gainsay the need for consultation within organisations or for lines of responsibility to funding sources to be maintained. Successful action needs decision, sometimes instant decision, but most of all inspired decision, a process not always favoured by the complex structures that have appeared and sometimes disappeared for the organisation of agricultural science in the 1970s and 1980s.

The way forward

One of the difficulties of organising science in the service of agriculture lies in the different motivation of research scientists, developers and entrepreneurial farmers. They exist in a continuum with, at one end, scientists, who are motivated not by financial reward but by the competition for approbation by their peers. At the other end the entrepreneurial farmer is driven by entrepreneurial excitement and by the profit motive – or at least the need for financial survival – although not necessarily without interest in the scientific and technical problems and in good farming practice. Agricultural developers share elements of both, and often carry within themselves a frustrated desire to be fully-fledged farmers. Finding ways of motivating this range of personalities is difficult but must be done to make the system work.

One other reason for the industry's being capable of facilitating research and development was that it had an adequate number of well-trained technologists, not only in the advisory and support services, but also employed directly in the industry. Another reason was its well-trained technicians and operatives. The educational provision at all levels (university, national college, county college and via the Agricultural Training Board) was comprehensive. Some sections of the industry also achieved a considerable measure of self-help in the research and development field. The Pig Industry Development Association (P.I.D.A., levy-funded) through its post-graduate scholarships, travel scholarships and support of research and development, helped to sustain a competitive pig industry, while its successor, the Meat and Livestock Commission, widened horizons and played a prominent part in applied research and development. Similar examples can be given on the crop side with the Potato Marketing Board and Home Grown Cereals Authority.

It would be unusual to find all the human resources required for a specific development project within one organisation; the mechanism has not always existed to facilitate cooperation between organisations for specific projects. This often happens best where there is mutual friendship between individuals and where the organisations are in close enough proximity to make conversation and exchange of ideas easily possible. But even here, once an agreement has been reached it is necessary to write protocols that give a leadership position to someone who can call on a range of resources and staff. Agricultural administrators must appreciate the problem and avoid structural barriers that impede the symbiosis between individuals and organisations.

What of the future? The future of agriculture will be what the world community decides it should be. Whatever policies are adopted, it is clear that scientists and agriculturists can deliver the necessary improved technology. The problem for the immediate future is a global one and one shared with other industries. How does the world organise its production so that the world-wide population is gainfully employed and adequately fed (Holmes, 1986), while the rural population is maintained at a desirable level in each country? How are the natural advantages and disadvantages of geography to be taken into account? Can this be done while still maintaining the natural flora and fauna, avoiding global warming and disastrous pollution of other kinds?

The message of this book is that, despite imperfections in the organisation of agricultural research and development in the past 50 years, the scientists of the modern agricultural revolution in the United Kingdom did deliver what was required. If a sound policy for world agriculture can be developed there is no doubt that the science of tomorrow is even better fitted to facilitate its implementation.

Before a world policy evolves, the agriculture of individual countries will need to be put on a proper footing. In Britain and Europe there is the current need to move from a supported agriculture to an agriculture that can sell products in an unsupported market place. But for how long? The economic arguments for the abolition of subsidised agriculture are powerful (McCrone, 1962; Raeburn, 1958); the human problems resulting from the withdrawal of subsidies remain daunting. The immediate answer to the problem is to seek to add value to the local (now unprotected) product: hence entrepreneurial farmers seek to make cheese, to produce organically grown crops, to deliver home-made ice cream, to open farm leisure parks with ancient breeds, etc., etc. All these activities are being undertaken today. More of the same are probably needed; the removal of subsidies will be a powerful spur to further innovation in areas where a profit can be made. But there is a limit to the amount of ice cream and farmhouse cheese that can be eaten, or the number of leisure parks that can be visited, and the length of time useful profit margins can be maintained against increased internal competition is doubtful.

Alternatively, at a cost, it is possible to use farming to produce an environmentally satisfying countryside; to promote leisure farming for those in industrial employment; or to facilitate environmentally friendly farming in terms of energy and chemical fertilisers. Before that can be done the cost of such initiatives has to be computed and the political will for their achievement has to emerge. It must not be thought that no development work is in train to encourage innovative crop and animal production. There are several projects, for example the Flaxlin project of the Ministry of Agriculture, which, following the E.E.C. subsidies for oil seed crops, has mounted an inter-disciplinary and inter-institutional approach to the utilisation of the fibre part of the linseed currently grown for oil.

One other possible route that is arousing interest arises from the increased capacity for genetic change in crop plants and animals made possible by molecular genetics and biotechnology. It involves the transformation of the genome in plants or animals, allowing: the production of crop plants adapted to a wider range of edaphic conditions than at present; the incorporation into crop plants of resistance to insect and plant pathogens; the production of specific substances (antibodies, blood components, enzymes, etc.) in quantity, in either plants or animals; the incorporation of specific qualities such as enhanced shelf life in tomatoes; and the production in quantity of chemical feedstocks for further use in manufacturing, including the modification of rape seed oil as a substitute for diesel fuel. These advances offer hope in two areas: firstly in world agriculture, where there is a prospect that problems such as the utilisation of saline soils and water sources and the adaptation of crops to resist water stress may be helped by the production of specially engineered plants. These problems are best taken care of by the large international research institutes with specialised help from scientists with particular skills. Once new crop varieties are identified there may be a role for the multinational seed houses but there must always be a problem in marketing varieties at a profit in developing countries. The second area of advantage is in the agriculture of advanced countries, where the dominating problems at the moment are economic; biotechnology offers the real but distant possibility of a range of new options for profitable enterprises. Of these the specialised husbandry of animals and plants modified to produce specific substances used in human medicine can form the basis of enterprises serving a small (even very small), but profitable, niche market. The other possibility of using crops as the source of a wider range of chemical feedstocks (cereals already form the feedstock for the production of alcohol for human consumption) as precursors of petrochemicals, for the internal combustion engine, and for textile and plastic precursors, etc., offers hope of a new agricultural revolution, where farmers are presented with an entirely new and reasonably stable market which at the same time presents technological challenges.

Large multinational companies are already looking at the possibilities and we can expect to see advances in this area driven by normal market forces in the foreseeable future. But is that sufficient? Some would answer in the affirmative and allow the normal economic forces to determine the range of new endeavours and where they are to take place. Others think that some direction can be given to this future and that the development process should be harnessed to determine what is needed, how far it can be provided, and by whom. New discoveries in biotechnology are tumbling out of the laboratories of the advanced countries, not least out of the laboratories of the former Agriculture and Food Research Council in Britain. Left entirely to the market, a number of these discoveries will be exploited to the best ability of commerce. Others will be passed over for reasons of company policy, or because another piece of research, which cannot be afforded, is required to make the innovation secure. Perhaps the development costs are too high; perhaps the company's product mix is perturbed by the new proposal; perhaps the cost of factory or plant modification is too high. The result may be that specific developments take place by chance in particular parts of the world and perhaps without releasing immediately their full potential.

The most difficult area to exploit is the production of substitute feedstocks for industry, because of the complex market forces that are at work (compare the energy industry and the competitive positions of coal, oil, gas, electricity and nuclear energy). However, if the focus is narrowed to look at Britain, it can be seen that there might be benefit in trying to ensure that some developments take place in conjunction with British agriculture. In an emerging situation, those who establish a process can often maintain commercial advantage just because of their position and know-how. If British agriculture and British industry can combine to ensure that a few at least of these developments take place in Britain, it will be to their mutual benefit. At the least a flourishing agriculture is a minimal charge on the taxpayer. For that reason government support for the basic research on molecular genetics for agriculture needs to be supplemented by funds to encourage the exploitation of the development of the emerging technology. It must all be set in a background of a new and enhanced political vision.

Bixler and Shemilt (1983) have articulated the need for exceptional support for several fundamental research projects, with the aim of lifting food production in disadvantaged areas by increments of 50%, 100% or even more, because of the severe shortfalls which will otherwise appear. An extension of this thinking suggests that an imaginative investment in one or two development projects, which might change the whole purpose of agriculture in Britain and spare it the need for support, would be sensible. If this involved the production of chemical feedstocks by the plant's photosynthetic energy there would be a bonus of energy saved and pollution prevented. Such development, the need for which is not easily recognised by businessmen or scientists, needs to be led by government policy, fiscal encouragement (to encourage industry to modify and update equipment to meet the new need; cf. agriculture in the 1970s) and by encouraging discussion and early starts to find out the possible avenues for further exploration. Moreover, such development work not only crosses the boundaries of several research councils and needs skilled organisation, but also functions in an industrial area where company interest demands confidentiality. It is a daunting prospect. However, without some such initiative we shall retain a conventional agriculture, which will shrink during the next period of industrial recovery and in the future will make a diminishing contribution to the wealth of the nation. The example of biotechnology and possible developments in the growing of industrial crops have been cited, but there will be other discoveries that may open other doors for agriculture. The point is that they must be looked for and exploited with vigour.

The preceding chapters have shown what science and individual scientists can do. Looking to the future, one of two courses can be followed. Either the present situation of a world agriculture with uneven production, areas of major production and areas of famine and deprivation, can be accepted; and at the same time we can shut our eyes to a world population with a doubling time of approximately 40 years and a possible population of 6 000 000 000 in the year 2000. In the same spirit, in Britain, a local agriculture with little future without support may be accepted. Alternatively, it is necessary to subscribe to what Blaxter (1986) has called 'technological optimism' and to fund and organise our scientific effort to conquer the problems of world starvation and, on quite a different scale, to give a new thrust and excitement to the British farming industry.

Throughout this chapter the importance of the correct organisational and funding framework for successful science and development for the agricultural industry has been stressed. Important it is; but most important of all, as was shown in the earlier chapters, are the men and women who have the ideas, the prescience, the energy and the persistence to carry the innovations to a successful conclusion. We must include here those 'awkward' scientists who do not fit within the organisational structure and who are nurturing the (as yet dimly perceived) science of tomorrow. Let us ensure that we give all the men and women of vision the opportunities and the environment to play their part.

Selected references and further reading

Bixler, G. and Shemilt, L.W. (1983). *Chemistry and World Food Supplies: The New Frontiers.* Pergamon Press, Toronto.

Blaxter, K.L. (1977). Energy and other inputs as constraints on food production. *Proceedings of the Nutrition Society* **36**, 267–73.

Blaxter, K.L. (1986). *People, Food and Resources.* Cambridge University Press.

Caird, J. (1878). *The Landed Interest and the Supply of Food.* Cassell, Petter and Galpin, London.

Fleming, I.F. and Robertson, N.F. (1990). *Britain's First Chair of Agriculture.* East of Scotland College of Agriculture, Edinburgh.

Henderson, W. G. (1981). British Agricultural Research and the Agricultural Research Council. In *Agricultural Research 1931-1981*, ed. G.W. Cooke, pp. 3–113. Agricultural Research Council, London.

Hill Farming Research Organisation. (1979). *Science and Hill Farming.* H.F.R.O., Edinburgh.

Holmes, J.C. (1986). Crop Production – Famine and Glut. Inaugural Lecture of the University of Edinburgh. In *Annual Report of the Edinburgh School of Agriculture.* East of Scotland College of Agriculture, West Mains Road, Edinburgh EH9 3JG, U.K.

McCann, N.F. (1989). *The Story of the National Agricultural Advisory Service – a Main Spring of Agricultural Revival*. Providence Press, 38b Station Road, Haddenham, Ely, U.K.

McCrone, G. (1962). *The Economics of Subsidising Agriculture*. Allen and Unwin, London.

Murray, K.A.H. (1955). *Agriculture* (History of the Second World War, U.K. Civil Series). H.M.S.O., London.

National Economic Development Office. (1985). *The Adoption of Technology, Opportunities for Improvement*. National Economic Development Office, Millbank Tower, London SW1P 4QX, U.K.

National Economic Development Office. (1986). *The Adoption of Technology*. I. *Improved Grassland Management in Hill Sheep Farming Systems*, II. *Silage*. National Economic Development Office, Millbank Tower, London SW1P 4QX, U.K.

Raeburn, J.R. (1958). Agricultural production and marketing. In *The Structure of British Industry* (two volumes), ed. D. Burn, pp. 1–46. National Institute of Economic and Social Research, Economic and Social Studies no. 15. Cambridge University Press.

Tracy, M. (1964). *Agriculture in Western Europe*. Jonathan Cape, London.

Winnifrith, J. (1962). *Ministry of Agriculture, Fisheries and Food*. Allen and Unwin, London.

Glossary

Anthesis. Stage of appearance of the anthers, particularly in the cereal flower.

Ark (chickens). A hut of triangular section to hold chickens for moving over grass.

Bacon weight (pig). The weight at which pigs are killed to make bacon (90–125 kg). This weight is heavier than pork weight.

Blastocyst. A hollow sphere of cells derived, by division, from the ovum.

Bolting. Premature flowering.

Broiler (chicken). A quickly grown chicken suitable for broiling (grilling) or more commonly roasting.

Catch crop. A crop taken between two main crops, often in the autumn.

Coccidiostat. A drug that reduces the multiplication rate of the protozoal coccidia.

Coulter. The part of a plough or seed drill which cuts a path through the soil for the ploughshare or seed delivery mechanism.

Combining pea. Pea crop grown for the production of dry peas to be harvested by grain combine.

Crimping. Pressing, pinching and indenting grass for hay (generally) to allow easier drying.

Day-neutral plant. A plant which flowers independently of changes in day length.

Dough stage (maize grain). The point at which the endosperm is full of starch but has not lost much water.

Ensilage. The process of making silage by allowing a limited fermentation of fresh herbage with the production of lactic and other acids as preservatives.

Entry (for crops). The point in the rotation at which the next crop can be sown conveniently.

Epiphytotic. An epidemic among plants.

Epizootic. An epidemic among animals.

Finished (animal). An animal fed and grown to its optimal economic weight.

Finishing period. The final period when an animal is being (well) fed preparatory to slaughter.

Finishing unit. A building or part of a building containing animals being fed for slaughter.

Flag leaf. The final leaf to emerge in a cereal before the appearance of the stem and ear.

Haulm. Above-ground stem of a plant (particularly of potatoes).

Heterosis. The (generally) increased vigour and production qualities of cross-bred domesticated plants and animals.

Horizontal resistance. Resistance shown by crops to plant pathogens, which is polygenically inherited, i.e. is not susceptible to sudden, single-step, genetic change.

Hysterectomy. The removal of the uterus, here from the gravid female.

Hysterotomy. The incision of the uterus to release the progeny (Caesarean section).

Lean content (meat). The proportion of edible protein in the carcase.

276

Ley. A temporary grass sward.

Ley farming. Farming in which intensive arable cropping alternates with a period in grass in the same field.

Lodging. Situation where a standing crop of cereals falls over because of cultivation errors, weather or disease.

Long-day plant. A plant that produces its flowers in response to increasing day length.

Malting barley. Barley grown so that the endosperm of the grain is low in nitrogen and suitable for the production of clear malt extract.

Monovalent vaccine. A vaccine active against a single organism or strain of organism.

Morula. A solid spherical mass of cells derived from the early division of the fertilised egg.

Periparturient period. The period that includes birth (and a few days either side).

Poly-tunnel. A simple protective structure (glasshouse substitute) made from metal hoops covered by polythene or equivalent.

Post- /pre-emergence weedkillers. Weedkillers that act on weeds, with safety to the crop, either before or after emergence of the sown crop.

Rogueing. Removing healthy plants of the wrong variety or type, or unhealthy plants from a crop grown for seed.

Self-mulching soil. Soil which, under minimal or no cultivation, develops a useful tilth from physical changes (produced by the weather) from the action of earthworms and the incorporation of organic matter.

Shatter. The break-up of the ripe ear or other fruiting structure before or during harvest, with consequent loss of yield.

Short-day plant. A plant that produces its flowers in response to decreasing day length.

Silo. A structure for containing silage and sometimes grain.

Singling. Removing excess plants from a sown drill so that remaining individual plants are separated from one another.

Stook. A group of sheaves (large bunches of harvested straw and grain) set to support one another in the stubble field.

Strip-grazing. Method whereby stock are allowed access to fresh and untrodden grass protected by an electric fence, moved periodically (often daily).

Swathe. Cut grass or oil seed rape gathered and laid in rows behind the mowing machine.

Tedding. Process of mechanically shaking up drying hay to expose it to the sun and air.

Tillers. The secondary shoots on a cereal plant, usually carrying ears themselves.

Tupping. Mating ewes with a ram (tup).

Unthriftiness. Failure to thrive, through illness or poor nutrition.

Vertical resistance. Resistance of plants to disease conferred by a single gene or genes and subject to rapid change.

Vining pea. Peas grown for the fresh market and harvested by a machine, which strips the pods and shells the peas.

Winter-proud (crops). Crops that have grown tall in a mild autumn and are susceptible to weather damage and to attack by (for example) rust disease.

Index

Note: page numbers in *italics* refer to tables